THE EVERYDAY SCIENCE

Where Curiosity and Reality Converge

Shah Rukh

CONTENTS

INTRODUCTION

Welcome to the captivating world of "The Everyday Science: Where Curiosity and Reality Converge." In the pages that follow, we embark on a journey of discovery that bridges the gap between the wonder of curiosity and the intricacies of reality. Science, often perceived as a realm of lab coats and equations, is, in fact, an integral part of our daily lives. It is the force that drives our understanding of the universe, from the mysteries of the cosmos to the mechanics of a simple everyday task.

This book is a celebration of the scientific wonders that surround us—a testament to the intricate web of knowledge that makes our world both comprehensible and endlessly fascinating. Whether you're an avid science enthusiast or someone seeking to unravel the mysteries of the everyday, this book is your guide through the intricacies of our existence.

From the vibrant colors that dance before our eyes to the hidden forces that shape our world, from the microscopic realm of cells to the mind-bending mysteries of quantum physics, "The Everyday Science" brings together the diverse tapestry of scientific disciplines to offer a comprehensive and accessible exploration of our reality. Through engaging narratives, illuminating explanations, and thought-provoking anecdotes, we dive into the core of what makes our world tick.

Prepare to be captivated by the magic of water as we delve into its essential role in shaping life, to marvel at the intricate dance of elements that compose the world around us, and to be awed by the remarkable impact of microbes on everything from our health to our environment. We'll uncover the secrets of colors and light, demystify the forces that guide the motion of our planet, and embark on cosmic adventures to the stars and beyond.

But this journey isn't just about the mechanics of the universe— it's about the human spirit of curiosity that propels us to explore, understand, and innovate. It's about the questions that arise as we gaze at the night sky, the amazement that surges within us when we witness the power of electricity, and the sense of wonder that accompanies our discoveries about the building blocks of matter.

As we journey through the pages of this book, we'll encounter the ever-present interplay of nature's teamwork, the dynamic forces of weather, the transformative energy that propels life, and the healing embrace of medical science. We'll explore the intricate journey of food from farm to fork, the materials that shape our world, and the astonishing workings of our brain—the seat of our consciousness and creativity.

Every chapter in "The Everyday Science" is a portal to deeper understanding, offering insights into the profound interconnectedness of all things and celebrating the ceaseless pursuit of knowledge that defines human civilization. From the wonders of the oceans to the mysteries of the quantum realm, from the marvels of genetics to the enigmas of time and space, we invite you to embark on this journey of exploration and curiosity. As we venture together through the realms of science, let us embrace the beauty of our world, the power of our curiosity, and the endless possibilities that unfold when curiosity and reality converge.

CHAPTER 1: THE MAGIC OF WATER: HOW IT SHAPES LIFE

Water, the essence of life, possesses an undeniable magic that shapes the very fabric of existence on Earth. From the serene flow of a meandering river to the crashing power of ocean waves, water's influence is both gentle and formidable, intricately woven into every facet of our planet's ecosystem. This intricate dance of hydrology is a story of connection, adaptation, and transformation that spans millions of years, giving rise to the extraordinary diversity of life that inhabits our blue planet.

The Primordial Playground: Origins of Water's Magic

Water's magic begins with its extraordinary molecular structure. H_2O, two hydrogen atoms bonded to a single oxygen atom, creates a unique molecule with remarkable properties. Its bent shape and polarity lead to hydrogen bonding, which imparts water with its cohesive and adhesive characteristics. This molecular arrangement contributes to water's high heat capacity, allowing it to absorb and release heat slowly, thus moderating temperature fluctuations on Earth.

Water's journey begins eons ago, as the Earth cooled and formed oceans, lakes, and rivers. These bodies of water became the cradle for the earliest forms of life, providing a nurturing environment where complex chemical reactions could take place. In these aquatic realms, the first single-celled organisms emerged, harnessing the power of water to catalyze reactions and eventually pave the way for the evolution of more advanced life forms.

The Elixir of Life: Water's Biological Significance

Water is the elixir of life. It composes a significant portion of all living organisms, enabling essential biological processes to unfold. Cells are bathed in an aqueous environment, facilitating nutrient transport, waste removal, and chemical reactions necessary for growth and reproduction. Plants, for instance, use water's capillary action to transport nutrients from roots to leaves, nourishing the entire organism.

Water's ability to dissolve a wide range of substances further enhances its biological significance. It acts as a universal solvent, facilitating the transport of minerals and nutrients through the environment and enabling organisms to extract the resources they need. This dissolved substance transport also plays a critical role in the Earth's carbon cycle, aiding the absorption of carbon dioxide by oceans and ultimately affecting the planet's climate.

A Symphony of Adaptation: Water's Influence on Life

Life has adapted to water's various forms and states, be it liquid, solid, or vapor. In colder environments, organisms have evolved mechanisms to survive freezing temperatures by producing antifreeze proteins that prevent ice crystal formation within their cells. In desert ecosystems, plants and animals have developed water-efficient strategies like storing water, minimizing water loss through specialized structures, and adjusting their metabolic rates to conserve moisture.

Aquatic life showcases an incredible array of adaptations as well. Marine animals have evolved streamlined shapes to navigate the water with minimal resistance, and deep-sea creatures have adapted to extreme pressure and darkness. Coral reefs, often referred to as the rainforests of the ocean, thrive in warm, nutrient-rich waters, forming complex ecosystems that support an astonishing diversity of marine life.

Water's Transformative Power: Shaping the Earth's Landscape

Water is an artist, sculpting the Earth's surface over millions of years. The erosive power of water has formed awe-inspiring landscapes, from the grandeur of the Grand Canyon to the delicate formations of slot canyons. Rivers carve through rock, creating valleys and gorges, while glaciers reshape mountains and valleys with their slow but relentless movement.

Water is also an architect of habitable environments. Through the shaping of coastlines, estuaries, and wetlands, it creates biodiversity hotspots and acts as a buffer against storms and erosion. Estuaries, where freshwater rivers meet the saltwater of oceans, are rich in nutrients and provide critical nurseries for numerous species, supporting both aquatic and terrestrial life.

The Dance of Sustenance: Water's Role in Food Security

Water's role in shaping life extends to global food security. Agriculture, the cornerstone of human civilization, is deeply dependent on water availability. Rivers and irrigation systems enable the cultivation of crops that sustain human populations, while aquaculture leverages water bodies to produce seafood, addressing the nutritional needs of millions.

However, this dance of sustenance is not without challenges. Water scarcity, pollution, and inefficient management threaten both agricultural productivity and the delicate balance of aquatic ecosystems. Climate change introduces further complexity, altering precipitation patterns, increasing the frequency of extreme weather events, and impacting water availability, thus underscoring the importance of responsible water resource management.

Conclusion: Water's Enduring Enchantment

In the grand tapestry of life on Earth, water's magic is undeniable and pervasive. It shapes landscapes, fuels biological processes, and serves as a cradle for evolution. Its molecular dance facilitates interactions at the cellular level and nurtures complex ecosystems. From the smallest microorganism to the vast expanse of oceans, water's influence is a reminder of the interconnectedness of all life.

Understanding and appreciating the magic of water is not only a scientific endeavor but also a philosophical and ethical one. As stewards of this planet, it's crucial to acknowledge water's vital role in shaping life and to work collectively to preserve its integrity. By doing so, we honor the intricate dance of water and life that has been unfolding for millennia, ensuring its continuation for generations to come.

CHAPTER 2: BUILDING BLOCKS OF STUFF: ELEMENTS AND HOW THEY STICK TOGETHER

At the heart of the intricate tapestry of matter that makes up the universe lies a captivating story of elements, the fundamental building blocks of all substances. These elements, with their unique properties and behaviors, come together through various bonding mechanisms to create the diverse materials that compose the world around us. Exploring the realm of elements and their interactions unveils the marvels of chemistry, from the simplicity of hydrogen to the complexity of molecules.

The Elemental Pantheon: A Periodic Table Odyssey

The periodic table, a cornerstone of modern chemistry, is a visual representation of the elements arranged in order of increasing atomic number. The table showcases the 118 known elements, each characterized by the number of protons in its nucleus. Hydrogen, the simplest element, possesses only one proton, while uranium, at the other end of the spectrum, boasts 92 protons.

Elements are classified into groups based on their chemical and physical properties. These groups, such as metals, non-metals, and metalloids, offer insights into how elements interact with each other and form compounds. Noble gases, for instance, are known for their inertness due to their stable electron configurations, while alkali metals are highly reactive due to their tendency to lose one electron.

Orbiting Electrons: The Glue that Holds Elements Together

At the heart of an element's behavior lies its atomic structure, composed of protons, neutrons, and electrons. Electrons, with their negative charge, orbit the nucleus in energy levels or shells. The arrangement of these electrons influences an element's chemical properties, especially its reactivity.

Chemical bonding, the process through which elements combine to form compounds, is governed by the interactions of electrons. Covalent bonding involves the sharing of electrons between atoms, leading to the formation of molecules. For instance, in a water molecule, two hydrogen atoms share electrons with an oxygen atom, creating a stable molecule with distinct properties.

Ionic Bonds: Electrons in Transit

Ionic bonding occurs when atoms transfer electrons from one to another. This results in the formation of ions, charged particles with unequal numbers of protons and electrons. Positively charged ions (cations) and negatively charged ions (anions) are drawn together by electrostatic forces, creating ionic compounds with specific crystalline structures. Sodium chloride (table salt) is a classic example of an ionic compound, where sodium donates an electron to chlorine.

Molecular Dance: Van der Waals Forces

Van der Waals forces, a type of intermolecular force, play a subtle yet significant role in bonding. These forces arise from temporary fluctuations in electron distribution, causing momentary charges. Though individually weak, the cumulative effect of these forces is crucial in holding molecules together. This is exemplified in the unique properties of water, where hydrogen bonds—an extreme form of dipole-dipole interaction—lead to high surface tension, capillary action, and anomalous expansion upon freezing.

Metals: Sea of Electrons

Metals, with their lustrous appearance and exceptional conductivity, possess a distinct bonding mechanism. In a metal, atoms are arranged in a closely packed lattice, with a "sea" of loosely held electrons freely moving between them. This "electron sea" facilitates thermal and electrical conductivity, making metals essential in countless applications, from wiring to construction.

Polymer Chains: Long Molecular Marathons

Polymers, large molecules composed of repeating subunits called monomers, showcase another fascinating bonding phenomenon. Covalent bonds link these monomers into long chains or networks, resulting in materials with diverse properties. From the flexibility of rubber to the rigidity of plastics, polymers offer versatility and adaptability.

Beyond Elements: Complexity Through Compounds

Compounds, formed through the combination of elements in specific ratios, take the complexity of chemistry to new heights. Each compound has its own set of properties, distinct from its constituent elements. Carbon dioxide, composed of one carbon atom and two oxygen atoms, demonstrates how the same elements in different combinations can yield diverse outcomes—a vital concept in understanding the complex chemistry of life.

Synthesis and Creativity: The Human Touch in Chemistry

The exploration of elements and their interactions is not only a scientific endeavor but also an artistic and creative one. Humans have harnessed the knowledge of chemistry to create an astonishing array of materials and compounds, from medicines that save lives to innovative materials that shape technology and architecture.

Conclusion: Elements in Harmony

The story of elements and their interactions is a tale of harmony and diversity. Elements, like notes in a symphony, come together in intricate arrangements to create the melody of the universe. Their behaviors, from simple covalent bonds to the complex dance of electrons, shape the materials that surround us. By understanding the nature of elements and their bonding mechanisms, we unravel the blueprint of matter and gain insights into the awe-inspiring complexity of the world we inhabit.

CHAPTER 3: TINY HELPERS: HOW MICROBES AFFECT EVERYTHING

In the seemingly invisible realms that escape our naked eye, a universe of life thrives. Microbes, those microscopic organisms including bacteria, archaea, fungi, viruses, and more, constitute the foundation of life's diversity and play an integral role in shaping the world we know. From the intricacies of our bodies to the health of ecosystems and the global climate, these tiny helpers wield significant influence over every facet of existence.

Microbes Unveiled: A Hidden World Unveiled

Microbes, the oldest life forms on Earth, have been present for billions of years. They inhabit nearly every environment imaginable, from the depths of the ocean to the soil beneath our feet. Often existing in inconceivable numbers, these organisms collectively outweigh all plants and animals on the planet. Their astonishing diversity, with countless species and functions yet to be discovered, paints a picture of a microbial realm teeming with complexity.

The Human Microbiome: Our Internal Ecosystem

Within our bodies resides an ecosystem of microbes known as the human microbiome. Comprising trillions of microbes, this internal community exerts profound effects on our health. In the gut, for instance, microbes aid in digestion, produce vitamins, and even modulate our immune system. Recent research has linked the gut microbiome to various health conditions, including obesity, autoimmune diseases, and mental health disorders, highlighting the intricate interplay between microbes and human well-being.

Microbes as Engineers: Bioremediation and Beyond

Microbes exhibit remarkable capabilities as environmental

engineers. They are essential players in bioremediation, a process that employs microbes to break down pollutants and contaminants. Microbes can transform toxic substances into harmless compounds, contributing to the restoration of ecosystems tainted by pollution. This microbial prowess also extends to wastewater treatment, where these tiny helpers efficiently remove organic matter and pollutants from water sources.

Agricultural Allies: Microbes and Plant Health

In agriculture, microbes prove to be invaluable allies. Certain microbes form symbiotic relationships with plants, enhancing nutrient uptake and aiding in stress resistance. Mycorrhizal fungi, for example, extend the reach of plant roots, enabling them to access nutrients otherwise out of reach. Additionally, some microbes act as natural pest controllers by targeting harmful insects, reducing the need for chemical pesticides.

Global Elemental Cycles: Microbes as Biogeochemical Players

Microbes are key players in the global cycling of essential elements like carbon, nitrogen, and sulfur. Through processes such as photosynthesis, respiration, and nitrogen fixation, microbes shape the availability and distribution of these elements. Cyanobacteria, for instance, are responsible for a substantial portion of Earth's oxygen production through photosynthesis, while nitrogen-fixing bacteria convert atmospheric nitrogen into forms that plants can use.

Microbes in Climate Regulation: The Carbon Connection

Microbes have a profound impact on the Earth's climate. The carbon cycle, driven by microbial activity, involves the uptake and release of carbon dioxide. Ocean-dwelling microbes, collectively known as phytoplankton, capture carbon dioxide through photosynthesis, transferring it to the deep ocean as they die and sink. On land, soil microbes play a role in decomposing organic matter and releasing carbon back into the atmosphere as carbon dioxide or methane, potent greenhouse gases.

Microbes and Beyond: Space Exploration and Medicine

The influence of microbes extends beyond our planet. In space exploration, understanding microbial behavior is critical to ensuring the health of astronauts and the success of long-duration missions. Moreover, microbes have found applications in medicine, where probiotics and microbial therapies are being explored to treat conditions ranging from gastrointestinal disorders to skin ailments.

Ethical Considerations: Harnessing Microbial Power Responsibly

While the potential of microbes is awe-inspiring, their power must be wielded responsibly. Manipulating microbial communities can have unintended consequences, such as disrupting ecosystems or triggering antibiotic resistance. As we tap into the capabilities of microbes for biotechnology, biofuel production, and more, ethical considerations must guide our actions to preserve the delicate balance of ecosystems and ensure a sustainable future.

Conclusion: Guardians of a Microbial Universe

Microbes, often overlooked due to their size, are truly the guardians of a microbial universe that influences everything around us. From human health to environmental resilience, their impact is profound and pervasive. As we continue to unveil the mysteries of these tiny helpers, we uncover the interconnectedness of life on Earth, underscoring the delicate balance that sustains our planet and its myriad inhabitants.

CHAPTER 4: COLORS AND LIGHT: WHAT MAKES THINGS SHINE

In the mesmerizing dance between light and matter, the world is painted with an array of colors that captivate our senses. The interplay between light waves and the materials they encounter gives rise to the vibrant hues that adorn our surroundings. Understanding the intricate mechanisms behind colors and the behavior of light unveils the magic of optics, revealing how the world around us shines in a spectrum of captivating shades.

The Spectrum of Light: A Vibrant Symphony

Visible light, the narrow sliver of the electromagnetic spectrum that our eyes can perceive, spans from the deep violet to the rich red. This spectrum is composed of waves of varying lengths and frequencies, with each color corresponding to a specific wavelength. The colors of a rainbow—red, orange, yellow, green, blue, indigo, and violet—are a testament to the dispersion of light, a phenomenon that occurs when light passes through a medium, like raindrops, and is separated into its constituent colors.

Reflection and Absorption: The Color of Surfaces

The colors we perceive are a result of how objects interact with light. When light strikes a surface, three scenarios can unfold: absorption, reflection, and transmission. Objects appear a certain color because they absorb certain wavelengths of light while reflecting others. For example, a red apple appears red because it absorbs all colors of light except red, which it reflects.

Pigments and Dyes: Mastering Color Manipulation

Artists and technologists have long harnessed the power of pigments and dyes to manipulate colors. Pigments are finely ground particles that scatter and absorb light to create specific colors. Mixing pigments enables the creation of a wide range of hues. Dyes, on the other hand, are molecules that absorb specific wavelengths of light and transmit others. The textile industry, for instance, relies on dyes to achieve a kaleidoscope of colors in fabrics.

Additive and Subtractive Color Mixing: A Delicate Balance

The process of color mixing—combining different colors to create new ones—varies depending on whether it's additive or subtractive. Additive color mixing involves combining colored lights, as seen in digital displays. The primary colors in additive mixing—red, green, and blue—create white when combined at full intensity. In subtractive color mixing, used in processes like painting and printing, colors are created by selectively absorbing or subtracting certain wavelengths of light. The primary colors in subtractive mixing—cyan, magenta, and yellow—produce black when mixed at full intensity.

Prisms and Rainbows: Unveiling Light's True Nature

Prisms are optical instruments that demonstrate the phenomenon of dispersion, revealing the hidden spectrum within white light. When light passes through a prism, it bends due to differences in its speed within the material. This bending causes different colors to refract by different amounts, creating a spread of colors similar to a rainbow. The iconic experiment conducted by Isaac Newton using a prism showcased how white light is composed of a spectrum of colors.

Color Perception and the Human Eye: A Neurological Marvel

Color perception is a neurological marvel that involves the coordination of the eyes, brain, and specialized cells called cones in the retina. Cones are sensitive to different wavelengths of light, with some responding best to red, others to green, and the rest to blue. The brain processes the signals from these cones to create the colorful images we perceive. Color blindness, a condition that affects some individuals, results from anomalies in the cones' sensitivity.

The Science of Color in Industries: Aesthetic and Functional Aspects

Colors play a pivotal role in various industries beyond art and design. Marketing and branding heavily rely on color psychology to evoke emotions and influence consumer behavior. The automotive industry carefully selects paint colors to reflect trends and customer preferences. In the realm of medicine, color-coded systems help differentiate medications and indicate health hazards.

Quantum Mechanics and Color: A Deep Dive

Color and light also have a deep connection with the realm of quantum mechanics. The interaction between light and matter is governed by quantum processes, such as the absorption and emission of photons by atoms. The energy levels of electrons in atoms determine the specific wavelengths of light that are absorbed and emitted, resulting in the unique colors we see.

Conclusion: A Spectrum of Wonders

The interplay between colors and light is a symphony of physics, biology, and perception that adds depth, beauty, and meaning to our world. From the iridescence of butterfly wings to the technicolor displays of digital screens, the science of colors and light enriches our understanding of the universe and allows us to explore the very essence of how things shine, revealing the astonishing complexity of nature's palette.

CHAPTER 5: WHY THINGS FALL: EXPLORING GRAVITY'S PULL

In the grand tapestry of the cosmos, an invisible force weaves its influence across vast distances, orchestrating the movement of planets, stars, and even the simple act of an apple falling from a tree. This force, known as gravity, is one of the fundamental cornerstones of physics, shaping the behavior of objects on Earth and beyond. Delving into the intricacies of gravity unveils its far-reaching impact on the universe and our everyday lives.

The Universal Tug: Understanding Gravity's Basics

Gravity is a force of attraction between objects with mass. Every object, regardless of its size, exerts a gravitational pull-on other object. This force is proportional to the masses of the objects involved and inversely proportional to the square of the distance between them. Sir Isaac Newton's law of universal gravitation mathematically describes this relationship, enabling scientists to predict how objects interact gravitationally.

Galactic Choreography: Gravity on Cosmic Scales

Gravity's influence extends far beyond our terrestrial realm. On a cosmic scale, it orchestrates the dance of celestial bodies. Planets orbit stars, moons orbit planets, and galaxies cluster together—all under the guidance of gravity's invisible hand. The force of gravity between Earth and the Moon, for instance, is what keeps the Moon in a stable orbit around our planet, shaping the tides and influencing Earth's rotation.

The Apple of Understanding: Gravity and the Theory of Falling

The iconic story of Isaac Newton's observation of a falling apple is emblematic of his exploration of gravity. Newton realized that the same force pulling the apple to the ground was responsible for keeping the Moon in its orbit and governing the motion of planets.

This revelation led to the formulation of his laws of motion and the law of universal gravitation, which laid the foundation for classical physics.

Weight and Mass: The Earthly Consequences of Gravity

While mass is a measure of the amount of matter in an object, weight is the force exerted on an object due to gravity. Weight varies with location, as the strength of the gravitational field changes. For example, an object on the Moon would weigh less than it does on Earth due to the Moon's weaker gravity. Mass, however, remains constant regardless of location.

Free Fall: The Equalizing Effect of Gravity

In the absence of air resistance, all objects, regardless of their mass, fall at the same rate when subjected to gravity. This principle, famously demonstrated by Galileo dropping objects from the Leaning Tower of Pisa (though possibly a myth), underscores the universality of gravitational acceleration. In a vacuum, a feather and a hammer would hit the ground simultaneously, defying our everyday intuition.

The Curvature of Space-Time: Einstein's General Theory of Relativity

While Newton's law of universal gravitation provides an accurate description of gravity in most situations, Albert Einstein's general theory of relativity introduced a new perspective. According to Einstein's theory, massive objects like planets and stars warp the fabric of space-time around them. Objects then move along paths dictated by the curvature of this space-time, resulting in the observed gravitational effects.

Gravity's Cosmic Mysteries: Dark Matter and Dark Energy

Despite its profound influence, much about gravity remains mysterious. The universe appears to contain more matter than can be accounted for by visible objects, leading to the hypothesis of dark matter—an elusive form of matter that doesn't emit light but exerts gravitational influence. Similarly enigmatic is dark energy, which appears to be driving the accelerated expansion of the universe.

Applications of Gravity: From Space Exploration to Earthly Engineering

Gravity's effects shape many aspects of modern life. In space exploration, an understanding of gravitational forces is crucial for plotting trajectories and navigating spacecraft. On Earth, engineers consider gravity when designing structures, vehicles, and even amusement park rides. The study of gravity has also contributed to technological advancements, such as GPS systems that rely on precise measurements of gravitational effects.

Conclusion: Gravity's Unwavering Embrace

Gravity's pull is an ever-present force that shapes the cosmos and molds our reality. It guides the orbits of planets, causes objects to fall, and influences the evolution of galaxies. From the tiniest particle to the largest celestial body, gravity's embrace is a testament to the intricate interconnectedness of the universe. As we continue to explore its mysteries, we unlock deeper insights into the fabric of space-time and gain a greater appreciation for the intricate dance of forces that govern our existence.

CHAPTER 6: SPARKING WIRES AND MAGNETS: THE WORLD OF ELECTRICITY

In the heart of the modern age, a phenomenon that once fascinated and perplexed ancient civilizations have transformed into one of the foundational pillars of human progress. Electricity, with its ability to power our homes, drive technological innovations, and illuminate our lives, has revolutionized the way we interact with the world around us. This journey into the electrifying world of sparking wires and magnets unveils the science, history, and transformative power of electricity.

Unveiling Electricity: A Historical Journey

Electricity, derived from the Greek word "elektron" meaning amber, was first observed in antiquity. The ancient Greeks discovered that rubbing amber against fur generated a mysterious force that could attract lightweight objects. However, it wasn't until the 17th and 18th centuries that electricity's secrets began to unravel. Benjamin Franklin's groundbreaking experiments with lightning and Leyden jars laid the foundation for understanding electrical phenomena.

Charge and Current: The Fundamental Concepts

Electricity revolves around the concept of electric charge—the property that some particles possess, either as positive or negative charges. Opposite charges attract, and like charges repel. When charges are in motion, they create an electric current. Electric current, measured in amperes (amps), is the flow of charged particles, typically electrons, through a conductor.

Conductors and Insulators: Paths and Barriers

Conductors are materials that allow electric charges to flow freely through them. Metals, due to their abundant free electrons, are excellent conductors. Insulators, on the other hand, inhibit the flow of electric charges. Rubber, plastic, and wood are common insulators. This distinction is fundamental to the functioning of electrical circuits and systems.

Ohm's Law: The Relationship Between Voltage, Current, and Resistance

Ohm's Law, formulated by Georg Simon Ohm, establishes the relationship between voltage, current, and resistance in an electrical circuit. It states that the current passing through a conductor between two points is directly proportional to the voltage across the two points and inversely proportional to the resistance of the conductor.

Electromagnetism: Where Wires Meet Magnets

The interaction between electricity and magnetism is a fundamental aspect of electromagnetism. When a current flows through a wire, it generates a magnetic field around the wire. Conversely, a moving magnetic field can induce an electric current in a nearby wire. This principle forms the basis of electromagnetic devices like generators and transformers.

Direct Current (DC) and Alternating Current (AC): Powering Our Lives

Electricity comes in two primary forms: direct current (DC) and alternating current (AC). DC flows in one direction, making it suitable for applications like batteries and electronic devices. AC, which reverses direction periodically, is the type of electricity used in most power grids and household appliances. The development of AC power transmission by Nikola Tesla and the ensuing "War of the Currents" with Thomas Edison played a pivotal role in shaping our modern electrical systems.

The Power Grid: Energy's Journey to Our Homes

Electricity is generated at power plants through various means such as coal, natural gas, nuclear reactions, and renewable sources like solar and wind. It then travels through a complex network of power lines, substations, and transformers, known as the power grid, before reaching our homes and businesses. The grid's design and maintenance are critical to ensuring reliable energy supply.

Electrical Safety: Managing a Powerful Force

While electricity has revolutionized our lives, it also poses inherent risks. Electrical safety is of paramount importance, as even small currents can cause harm. Grounding, using circuit breakers, and following proper procedures for working with electrical equipment are essential precautions to prevent electrical accidents.

Electronic Revolution: Miniaturization and Innovation

The electronic revolution, driven by advancements in semiconductor technology, has transformed our world. The creation of transistors, integrated circuits, and microprocessors enabled the miniaturization of electronic devices, paving the way for computers, smartphones, and a host of technologies that have reshaped communication, entertainment, and industry.

Conclusion: Lighting Up the Future

Electricity's journey from the ancient mysteries of amber rubbing to the bustling power grids of today's cities showcases the remarkable progression of human knowledge and innovation. It's a force that has illuminated our lives, powered our dreams, and connected us across distances. As we look to the future, the world of sparking wires and magnets continues to evolve, promising new discoveries and transformative breakthroughs that will shape the trajectory of human civilization for generations to come.

CHAPTER 7: INSIDE ATOMS: THE TINIEST PIECES OF MATTER

In the heart of the intricate fabric of the universe lies a realm so minuscule that it defies our senses and challenges our understanding. This realm, the domain of atoms, constitutes the building blocks of matter, shaping the world in ways both subtle and profound. Exploring the depths of atoms unveils a world of particles, forces, and energies that underpin the very essence of reality.

The Atom's Structure: A Quantum Puzzle

At the dawn of modern atomic theory, John Dalton's model depicted atoms as indivisible spheres, like tiny marbles. However, the emergence of quantum mechanics shattered this simplicity, revealing that atoms are composed of even smaller particles. The basic structure of an atom consists of a nucleus—a dense, positively charged core—and a cloud of negatively charged electrons orbiting around it.

Protons, Neutrons, and Nuclei: The Atom's Core

At the heart of an atom resides its nucleus, a remarkably dense and compact region. The nucleus contains protons, positively charged particles, and neutrons, which carry no charge. The number of protons in an atom determines its atomic number, defining its chemical identity. The arrangement of protons and neutrons within the nucleus dictates an atom's mass number.

Electrons: Dance of the Subatomic Particles

Electrons, with their negligible mass, orbit the nucleus in distinct energy levels or electron shells. Quantum mechanics introduced the concept of electron probability clouds, indicating the likelihood of finding an electron in a specific region around the nucleus. These shells are characterized by the number of electrons they can accommodate, with the innermost shells holding fewer electrons and the outer shells accommodating more.

Quantum Mechanics: The Rules of the Subatomic Realm

Quantum mechanics, the branch of physics governing the behavior of particles on an atomic and subatomic scale, presents a profound departure from classical physics. It introduced concepts such as wave-particle duality, where particles like electrons exhibit both particle-like and wave-like behaviors. Heisenberg's uncertainty principle posits that certain pairs of properties, like position and momentum, cannot be precisely known simultaneously.

Forces and Interactions: Glue of the Subatomic World

The subatomic realm is governed by four fundamental forces: gravity, electromagnetism, the strong nuclear force, and the weak nuclear force. Gravity and electromagnetism are familiar forces, while the strong nuclear force binds protons and neutrons within the nucleus. The weak nuclear force governs processes like radioactive decay.

The Standard Model: A Particle Symphony

The Standard Model of particle physics encapsulates our current understanding of the subatomic realm. It identifies a variety of particles, including quarks, leptons, and bosons, as the fundamental constituents of matter and carriers of forces. The Higgs boson, famously discovered at the Large Hadron Collider, is associated with endowing particles with mass.

Beyond the Atom: Subatomic Particles and Particles of the Universe

The subatomic realm extends beyond the atom, with particles like quarks and leptons serving as the building blocks of protons, neutrons, and electrons. Experiments in high-energy particle physics reveal the intricacies of these particles, shedding light on their properties and interactions. Moreover, the cosmos itself is composed of particles, such as neutrinos, that have fascinating properties and play significant roles in astrophysical phenomena.

Antimatter and Exotic Matter: The Flip Side of Particles

For every particle in the Standard Model, there exists an antimatter counterpart. Antiparticles possess opposite charges to their corresponding particles. When particles and their antiparticles collide, they annihilate, converting their mass into energy. This phenomenon has practical applications in fields like positron emission tomography (PET) scans in medicine.

Unraveling Mysteries: The Quest for Grand Unification

Despite the remarkable achievements of modern physics, many mysteries remain unsolved. Grand unification theories seek to unify the fundamental forces into a single, coherent framework. Additionally, the nature of dark matter and dark energy, which collectively make up most of the universe's content, remains elusive, hinting at new frontiers yet to be explored.

Conclusion: The Quantum Symphony of Reality

Peering into the atom is akin to gazing into the heart of the universe's symphony. The subatomic realm, governed by quantum rules, reveals a tapestry of particles, forces, and interactions that shape the very fabric of reality. As we journey deeper into the subatomic world, we continue to unveil its enigmatic nature, drawing closer to understanding the ultimate nature of matter, energy, and the cosmos itself.

CHAPTER 8: MUSIC AND NOISE: THE SCIENCE OF SOUND

In the vast symphony of existence, the phenomenon of sound weaves intricate melodies and rhythms that resonate with our emotions and perceptions. From the soothing harmonies of music to the chaotic clamor of noise, the science of sound unravels the principles that govern these auditory experiences. Delving into the intricate world of vibrations, frequencies, and perception, we explore the science that underpins the creation, transmission, and reception of sound.

Sound as a Wave: Vibrations in the Air

Sound is a form of energy that travels through a medium, most commonly air. It is generated by vibrations—oscillations of objects that create waves of pressure changes. These waves propagate as compressions and rarefactions, creating areas of high and low pressure in the medium. When these pressure waves reach our ears, they are transformed into the auditory sensations we perceive as sound.

Pitch and Frequency: The Musical Spectrum

Pitch, a fundamental characteristic of sound, corresponds to the perceived highness or lowness of a sound. This sensation is determined by the frequency of the sound wave—the number of oscillations per unit of time. High-frequency waves create high-pitched sounds, while low-frequency waves produce low-pitched sounds. Musical notes are categorized by specific frequencies and organized into scales to create harmonious melodies.

Timbre: The Color of Sound

Beyond pitch and loudness, sound possesses another quality called timbre, also known as tone color. Timbre accounts for the distinctive characteristics that differentiate sounds produced by different instruments or voices, even when they play the same note at the same volume. It is influenced by factors such as harmonic content, attack, and decay, which contribute to the richness and uniqueness of a sound.

Noise and Harmony: Dichotomy of Sound

Sound can be broadly classified into two categories: music and noise. Music is a deliberate arrangement of sounds with rhythm, melody, and harmony, often intended to convey emotions or express artistic ideas. Noise, on the other hand, lacks the intentional structure of music and is characterized by random, irregular vibrations that create dissonance. While music elicits aesthetic pleasure, noise can be disruptive and overwhelming.

The Doppler Effect: Shifting Frequencies

The Doppler effect is a phenomenon in which the perceived frequency of a sound wave changes due to relative motion between the source of the sound and the observer. As a source approaches, the waves are compressed, resulting in a higher perceived frequency (higher pitch). Conversely, as the source moves away, the waves are stretched, resulting in a lower perceived frequency (lower pitch). The Doppler effect is responsible for the familiar change in pitch of a siren as it passes by.

Echoes and Reverberation: Reflections of Sound

When sound waves encounter a barrier, they can be reflected, creating echoes. An echo is a delayed repetition of a sound due to its reflection off surfaces like walls or buildings. Reverberation, on the other hand, refers to the persistence of sound in an enclosed space due to multiple reflections. Reverberation is crucial in determining the acoustic properties of concert halls and recording studios.

Sound Perception: The Inner Ear and Brain

The process of sound perception begins with the ear capturing sound waves through the outer ear and channeling them to the eardrum. The middle ear then amplifies these vibrations and transmits them to the inner ear, where the cochlea—a spiral-shaped, fluid-filled structure—converts them into electrical signals. These signals are transmitted to the brain via the auditory nerve, where they are interpreted as sound.

Synesthesia and Emotional Impact: The Power of Sound

Sound not only engages our auditory sense but can also evoke emotional responses and even cross sensory boundaries in a phenomenon called synesthesia. Certain sounds can trigger memories, evoke nostalgia, or induce specific feelings. The use of music in films and other media is a testament to its ability to heighten emotions and enhance storytelling.

Technology and Acoustics: Shaping Sound Environments

Advances in technology have revolutionized the creation and manipulation of sound. Acoustics, the study of how sound behaves in various environments, plays a pivotal role in designing spaces with optimal sound quality. From concert halls engineered to enhance musical performances to noise-canceling headphones that attenuate unwanted sounds, technology has reshaped our interactions with sound.

Conclusion: The Sonic Tapestry of Life

Sound is more than just vibrations in the air—it's a fundamental aspect of our existence that resonates with our emotions, memories, and perceptions. Whether it's the soothing melodies of a lullaby, the energizing rhythms of a dance beat, or the cacophony of urban life, sound shapes our experiences in profound ways. The science of sound, encompassing physics, biology, psychology, and art, unravels the complexities of this sensory phenomenon, inviting us to explore the depths of its beauty and intricacy in the sonic tapestry of life.

CHAPTER 9: WHAT'S INSIDE CELLS? A PEEK INTO LIFE'S BLUEPRINT

In the quiet expanse of life, where mysteries of existence unfold, cells stand as the building blocks of all living things. These microscopic powerhouses harbor an astonishing array of structures and processes that govern the complexity and diversity of life on Earth. Exploring the inner workings of cells unveils the intricate machinery, genetic code, and dynamic interactions that define the blueprint of life.

The Cell: Foundation of Life

The cell is the basic unit of life. It's a dynamic microcosm that carries out vital functions essential for survival and growth. While cells vary in size, shape, and specialization, they share core features, such as a plasma membrane that separates them from their environment and genetic material that encodes the instructions for life processes.

Cell Membrane: Gateway to the Cell

The cell membrane, or plasma membrane, acts as a selective barrier that controls the passage of molecules in and out of the cell. Composed of a lipid bilayer embedded with proteins, the membrane maintains cellular integrity while facilitating communication and interactions with the external environment.

Organelles: Cellular Compartments

Within the cell's confines, various organelles perform specialized functions. The nucleus, often referred to as the cell's control center, houses the genetic material in the form of DNA. The endoplasmic reticulum is involved in protein synthesis, while the Golgi apparatus modifies, sorts, and packages proteins for transport. Mitochondria generate energy through cellular respiration, and lysosomes contain enzymes for digestion.

Cytoplasm and Cytoskeleton: Support and Structure

The cytoplasm, the jelly-like substance that fills the cell, provides a medium for chemical reactions and organelle movement. The cytoskeleton, a dynamic network of protein filaments, maintains cell shape, facilitates intracellular transport, and is crucial in cell division.

DNA and Genetic Code: Unraveling Life's Instructions

Deoxyribonucleic acid (DNA) holds the genetic code that dictates an organism's traits and functions. DNA's double-helix structure contains sequences of nucleotide bases—adenine, cytosine, guanine, and thymine—that encode information. The process of DNA replication ensures that genetic information is faithfully transmitted during cell division.

RNA and Protein Synthesis: From Code to Function

Ribonucleic acid (RNA) plays a key role in translating the genetic code into functional proteins. Transcription, the process of copying DNA into RNA, occurs in the nucleus. The resulting RNA molecules move to the cytoplasm, where translation converts the RNA code into a sequence of amino acids, forming proteins with diverse functions.

Cellular Energy: Powerhouses and Photosynthesis

Energy is essential for cell function, and two major processes provide it: cellular respiration and photosynthesis. Cellular respiration occurs in the mitochondria and involves breaking down glucose to produce energy-rich molecules like adenosine triphosphate (ATP). In photosynthesis, chloroplasts capture light energy to convert carbon dioxide and water into glucose and oxygen.

Cell Communication: Signaling Pathways

Cells communicate through intricate signaling pathways that coordinate responses to changes in their environment or internal state. Hormones, neurotransmitters, and growth factors trigger these pathways, often involving the activation of receptor proteins that relay messages to the cell's interior.

Cell Division: Growth and Renewal

Cell division is essential for growth, development, and repair. Two main types of cell division exist: mitosis and meiosis. Mitosis results in two genetically identical daughter cells, while meiosis leads to the formation of sex cells (gametes) with half the genetic material. The cell cycle, a series of stages involving growth, replication, and division, ensures the orderly progression of cell division.

Cell Diversity and Specialization: Unity in Variety

Cells exhibit diverse forms and functions based on their specialization. Stem cells, with their unique ability to differentiate into various cell types, contribute to the development and maintenance of tissues. Differentiation involves the activation of specific genes that lead to distinct cellular structures and functions.

Cancer and Cell Regulation: The Disruption of Balance

Cancer arises from the uncontrolled growth and division of cells. Mutations in genes that regulate the cell cycle, such as tumor suppressor genes and oncogenes, can lead to the loss of cell cycle control and the development of tumors. Understanding the genetic and molecular mechanisms of cancer is crucial for developing effective treatments.

Conclusion: The Tapestry of Life's Building Blocks

Within the hidden microcosm of cells lies the foundation of life's intricate tapestry. Cells, with their organelles, DNA, energy processes, and intricate signaling networks, create the symphony of life that spans from the microscopic to the macroscopic. As we unravel the secrets of cells, we gain insights into the fundamental processes that govern life's diversity, continuity, and evolution, underscoring the profound interconnectedness of all living things.

CHAPTER 10: PUSHING AND PULLING: FORCES AND BALANCE

In the grand theater of the universe, forces are the actors that dictate the motion and equilibrium of objects. From the graceful glide of a kite in the wind to the mighty pull of a planet on its moon, the concept of forces governs every aspect of motion and stability. By delving into the intricate interplay of pushing, pulling, and balance, we uncover the fundamental principles that shape the dynamic dance of the cosmos.

Forces: Unveiling the Invisible

Forces are interactions that cause objects to accelerate or deform. They can be divided into contact forces, which require direct physical contact between objects (like pushing a door), and non-contact forces, which act at a distance (like gravity). Forces are vectors, possessing magnitude and direction, and they are described using Newton's laws of motion—a cornerstone of classical physics.

Newton's Laws of Motion: A Blueprint for Dynamics

Sir Isaac Newton's laws of motion laid the foundation for understanding how forces influence motion. The first law, the law of inertia, states that an object will remain at rest or in uniform motion unless acted upon by an external force. The second law links force, mass, and acceleration: $F = ma$. The third law asserts that for every action, there is an equal and opposite reaction.

Balanced and Unbalanced Forces: Shaping Motion

When forces on an object cancel each other out, the object experiences a balanced force situation and remains at rest or in constant velocity. In contrast, unbalanced forces result in acceleration or changes in velocity. This principle is evident in the everyday experiences of pushing a shopping cart or kicking a soccer ball.

Friction: A Force of Opposition

Friction, a contact force, opposes the motion of objects sliding or rolling across surfaces. It's a vital force in everyday life, both hindering motion and enabling it. Different types of friction, such as static friction (preventing the initiation of motion) and kinetic friction (slowing objects in motion), play roles in everything from walking to driving.

Gravity: The Universal Attraction

Gravity is one of the most fundamental and pervasive forces in the universe. Every mass attracts every other mass with a force proportional to their masses and inversely proportional to the square of the distance between them (Newton's law of universal gravitation). Gravity shapes planetary motion, creates tides, and keeps objects grounded.

Electromagnetic Force: The Dance of Charged Particles

The electromagnetic force is responsible for the interactions between charged particles. It encompasses the attractive force between oppositely charged particles and the repulsive force between like charges. The electromagnetic force governs everything from the behavior of electrons around atomic nuclei to the intricate operations of electrical devices.

Balancing Act: Equilibrium and Stability

Equilibrium is a state in which the net force and net torque acting on an object are both zero. Objects in equilibrium may be at rest (static equilibrium) or moving at a constant velocity (dynamic equilibrium). Stability depends on the position of an object's center of gravity—the point where all the object's weight is concentrated.

Action and Reaction: Newton's Third Law in Action

Newton's third law of motion, often summarized as "for every action, there is an equal and opposite reaction," is at the heart of many interactions. When one object exerts a force on another, the second object exerts an equal and opposite force on the first. This interplay is what propels rockets, birds in flight, and athletes in action.

Forces in Nature: From Microscopic to Cosmic Scales

Forces extend beyond our immediate surroundings. In particle physics, forces at the quantum level, like the strong nuclear force that binds atomic nuclei and the weak nuclear force involved in certain types of radioactive decay, shape the behavior of subatomic particles. On cosmic scales, gravitational forces orchestrate the motion of galaxies and clusters of galaxies.

Engineering and Forces: Constructing the World

Understanding forces is essential in engineering, where the design of structures, machines, and devices relies on maintaining balance and managing stress and strain. Engineers employ concepts like statics (study of objects in equilibrium) and dynamics (study of forces causing motion) to create efficient and safe designs.

Conclusion: The Symphonic Web of Forces

Pushing and pulling, attracting and repelling—forces are the threads that weave the intricate tapestry of the universe. From the subatomic realm to celestial bodies, the interplay of forces shapes the dynamic choreography of the cosmos. By unraveling the principles of forces and balance, we uncover the secrets behind motion, stability, and the delicate equilibrium that sustains the universe's ceaseless dance.

CHAPTER 11: NATURE'S TEAMWORK: HOW LIVING THINGS DEPEND ON EACH OTHER

In the intricate web of life on Earth, every living organism is a thread interwoven with countless others, creating a tapestry of interdependence and coexistence. This intricate dance of relationships, known as ecological interactions, forms the backbone of ecosystems, shaping the balance, diversity, and sustainability of our planet's biosphere. Delving into the depths of nature's teamwork unveils a complex world where species collaborate, compete, and rely on each other for survival and flourishing.

Ecosystems: The Stage for Interdependence

Ecosystems are dynamic systems where living organisms—plants, animals, microorganisms—interact with each other and their environment. These interactions are fundamental to the flow of energy, cycling of nutrients, and the intricate balance that characterizes natural habitats, from towering rainforests to expansive oceans and arid deserts.

Symbiosis: Partnerships in Nature

Symbiosis refers to close and long-term interactions between different species, often resulting in benefits for one or both parties. There are three main types of symbiosis: mutualism (both species benefit), commensalism (one species benefits while the other is unaffected), and parasitism (one species benefits at the expense of the other). Examples of mutualism include pollination, where bees and flowers cooperate in a vital ecological dance, and mycorrhizal partnerships between plants and fungi, enhancing nutrient uptake.

Predation and Herbivory: The Circle of Life

Predation and herbivory are interactions where one species consumes another for sustenance. Predators hunt and feed on prey, shaping population dynamics and controlling species populations. Herbivores, in turn, graze on plants, influencing plant growth and distribution. These interactions create a balance in ecosystems, preventing any one species from dominating and ensuring biodiversity.

Competition: The Drive for Resources

Competition arises when species vie for limited resources like food, water, and shelter. This process shapes the distribution and abundance of species, leading to adaptations that allow them to occupy specific niches. Competitive exclusion principle suggests that two species with identical ecological needs cannot coexist indefinitely; one will eventually outcompete the other.

Trophic Levels: The Food Chain

Trophic levels represent the different stages of energy transfer in an ecosystem. Producers (plants) capture energy from the sun through photosynthesis, while consumers (animals) obtain energy by consuming other organisms. Primary consumers eat producers, secondary consumers eat primary consumers, and so on. Decomposers, such as bacteria and fungi, break down dead organic matter, returning nutrients to the ecosystem.

Keystone Species: The Architects of Ecosystems

Certain species, known as keystone species, have a disproportionately large impact on their environment compared to their abundance. Their presence or absence can significantly affect the balance of an ecosystem. For example, sea otters in kelp forest ecosystems prevent overgrazing by sea urchins, allowing kelp forests to thrive and support a diverse array of species.

Ecological Succession: Nature's Makeover

Ecological succession is the process by which ecosystems change and develop over time. It occurs through primary succession, where new habitats form on bare, lifeless surfaces, and secondary succession, where ecosystems recover after disturbances like fires or deforestation. Each stage of succession paves the way for new species to colonize and shape the ecosystem.

Human Impact: Unraveling the Tapestry

Human activities have profound effects on ecological interactions. Habitat destruction, pollution, overexploitation, and climate change disrupt ecosystems, leading to species extinctions, imbalanced food webs, and loss of biodiversity. Conservation efforts aim to mitigate these impacts and restore ecosystems to their natural state, highlighting the urgency of preserving nature's teamwork.

Conclusion: The Symphony of Life's Interconnections

Nature's teamwork is a symphony of collaboration, competition, and coexistence that forms the intricate relationships among living things. From the smallest microorganisms to the grandest ecosystems, these interactions shape the world around us, regulating vital processes, maintaining biodiversity, and nurturing the delicate equilibrium that sustains life on Earth. As we peer into the complex tapestry of nature's interdependence, we gain a deeper appreciation for the delicate choreography that unites all living beings in a harmonious dance of existence.

CHAPTER 12: SUN OR STORM: LEARNING ABOUT WEATHER

In the ever-changing tapestry of the atmosphere, weather orchestrates the dance of nature's elements. From the gentle caress of sunlight on a warm day to the tumultuous fury of a thunderstorm, weather shapes our lives, influences our activities, and connects us to the vast processes of Earth's atmosphere. Unraveling the intricacies of weather reveals the science behind atmospheric phenomena, the factors that drive them, and the fascinating interplay of forces that give rise to the atmospheric conditions we experience.

Atmosphere: Earth's Protective Blanket

The atmosphere, a gaseous envelope surrounding our planet, is a dynamic system composed mainly of nitrogen and oxygen. It is divided into distinct layers: the troposphere, stratosphere, mesosphere, thermosphere, and exosphere. These layers vary in composition, temperature, and pressure, playing a crucial role in shaping weather patterns and atmospheric phenomena.

Weather and Climate: Unraveling the Distinction

Weather and climate are interconnected but distinct concepts. Weather refers to the short-term atmospheric conditions in a specific area, including temperature, humidity, precipitation, wind speed, and atmospheric pressure. Climate, on the other hand, encompasses the long-term average of weather patterns in a region, providing insights into the seasonal and yearly trends that characterize a specific area.

Heat and Energy: The Sun's Influence

The sun is the ultimate source of energy that drives Earth's weather systems. Uneven heating of the Earth's surface by the sun's rays sets in motion the complex processes of convection, conduction, and radiation that give rise to atmospheric circulation, creating the winds and temperature gradients that influence weather patterns around the globe.

Pressure and Wind: The Atmosphere's Dynamic Dance

Differences in atmospheric pressure across regions lead to the movement of air, creating wind patterns. High-pressure systems are associated with descending, stable air and fair weather, while low-pressure systems involve ascending air, which often leads to cloud formation and precipitation. The Coriolis effect, caused by the Earth's rotation, deflects moving air masses, influencing the direction of winds.

Humidity and Cloud Formation: Water's Role

Water vapor, an essential component of the atmosphere, plays a pivotal role in weather processes. As air rises and cools, it loses its capacity to hold moisture, leading to condensation and cloud formation. Clouds, ranging from fluffy cumulus to towering cumulonimbus, offer visual clues about the atmospheric conditions and predict impending weather changes.

Precipitation and the Water Cycle: From Sky to Earth

Precipitation, in the form of rain, snow, sleet, or hail, is the culmination of atmospheric processes. Water evaporates from the Earth's surface, rises as vapor, condenses into clouds, and eventually falls back to the ground. The water cycle, which encompasses these processes, ensures the continuous movement of water through the atmosphere and the Earth's surface.

Severe Weather: Nature's Fury

Severe weather events, such as thunderstorms, tornadoes, hurricanes, and blizzards, captivate our attention with their intensity and power. These events arise from intricate interactions of temperature, humidity, and atmospheric instability. Thunderstorms spawn lightning, tornadoes result from powerful rotating updrafts, hurricanes draw energy from warm ocean waters, and blizzards bring fierce cold and snowfall.

Weather Forecasting: Deciphering Nature's Clues

Predicting weather conditions has come a long way from ancient methods based on observation. Modern meteorology employs sophisticated tools like weather satellites, radar systems, weather balloons, and computer models that simulate atmospheric processes. Meteorologists analyze data from these sources to create forecasts that aid in preparedness and decision-making for everything from daily plans to disaster response.

Climate Change: Earth's Changing Weather Patterns

The global climate is undergoing significant shifts due to anthropogenic activities, primarily the burning of fossil fuels that release greenhouse gases into the atmosphere. This results in the greenhouse effect, trapping heat and leading to rising global temperatures, altered weather patterns, more frequent extreme weather events, and shifts in precipitation patterns.

Conclusion: Embracing the Elements

Weather is a dynamic force that shapes the rhythm of our lives, influencing everything from agriculture to transportation, from clothing choices to our emotional well-being. Exploring the intricate mechanisms that drive weather patterns invites us to embrace the harmonious interplay of forces that govern our atmosphere. Whether we bask in the warmth of the sun or find shelter from a storm's fury, weather connects us to the larger symphony of Earth's intricate and ever-changing atmospheric orchestra.

CHAPTER 13: SHAKING EARTH: VOLCANOES, QUAKES, AND MOVING PLATES

Beneath the tranquil surface of our planet lies a dynamic realm where powerful forces shape the landscape and sculpt the very foundations of Earth. Volcanoes erupt in fiery displays of molten rock, earthquakes send shockwaves through the ground, and tectonic plates relentlessly shift and collide. This volatile dance, driven by the movement of Earth's lithospheric plates, unveils the geological processes that have shaped our world for millions of years.

Plate Tectonics: Earth's Jigsaw Puzzle

The theory of plate tectonics is the unifying framework that explains the movements and interactions of Earth's lithospheric plates. These plates, comprising the Earth's crust and a portion of the upper mantle, fit together like a complex jigsaw puzzle. They float on the semi-fluid asthenosphere beneath, driven by convection currents in the mantle.

Types of Plate Boundaries: The Contact Zones

Plate boundaries are regions where tectonic plates interact. There are three primary types of plate boundaries:

1. **Divergent Boundaries**: At divergent boundaries, plates move away from each other. This occurs at mid-ocean ridges, where magma wells up from below, creating new oceanic crust. An example is the Mid-Atlantic Ridge.

2. **Convergent Boundaries**: Convergent boundaries involve plates moving toward each other. One plate may be forced beneath the other in a process called subduction. This can lead to the formation of deep ocean trenches

and volcanic arcs. The collision of continental plates also creates mountain ranges, like the Himalayas.

3. **Transform Boundaries**: At transform boundaries, plates slide past each other horizontally. This lateral movement often results in earthquakes along faults. The San Andreas Fault in California is a well-known transform boundary.

Volcanic Activity: The Earth's Fiery Heart

Volcanoes are windows into Earth's molten interior. They form where magma, molten rock from the mantle, breaches the surface. This eruption can be explosive, as seen in stratovolcanoes like Mount St. Helens, or effusive, resulting in gently sloping shield volcanoes like Mauna Loa. Volcanic activity can also lead to the formation of new land, as lava cools and solidifies.

Earthquakes: Rumbling Beneath Our Feet

Earthquakes are the result of the sudden release of energy in the Earth's crust. Most earthquakes occur along faults, fractures in the Earth's crust where movement has occurred. The point within the Earth where the earthquake originates is called the focus, while the point directly above it on the Earth's surface is the epicenter. The Richter scale and the moment magnitude scale measure earthquake magnitude, quantifying their intensity.

Seismic Waves: The Tremors of Quakes

When an earthquake occurs, it generates seismic waves that radiate from the focus. There are three main types of seismic waves:

1. **Primary Waves (P-Waves)**: These are the fastest seismic waves, traveling through solids, liquids, and gases. They cause particles to move in the direction of the wave's motion.

2. **Secondary Waves (S-Waves)**: S-Waves move more slowly than P-Waves and only travel through solids. They cause particles to move perpendicular to the

wave's motion.

3. **Surface Waves**: These waves move along the Earth's surface and cause the most damage during an earthquake. They include Love waves, which move in a side-to-side motion, and Rayleigh waves, which create a rolling motion.

Impact on Society: Risks and Preparedness

Volcanic eruptions and earthquakes can have significant impacts on human populations and infrastructure. Volcanic ash can disrupt air travel and affect agriculture, while earthquakes can trigger tsunamis, landslides, and infrastructure damage. Understanding the geological processes and monitoring seismic activity allows societies to prepare and respond to these natural hazards.

The Ring of Fire: A Zone of Activity

The Pacific Ring of Fire is a horseshoe-shaped region encircling the Pacific Ocean known for its high tectonic activity. It's home to frequent earthquakes, numerous active volcanoes, and the subduction of several tectonic plates. The Ring of Fire serves as a vivid reminder of Earth's dynamic nature and the relentless forces that shape its surface.

Geological Time: Earth's Ever-Changing Story

The processes of plate tectonics, volcanic activity, and earthquakes are integral parts of Earth's geological history. They have shaped continents, raised mountains, sculpted ocean basins, and contributed to the evolution of life. The study of Earth's past helps scientists predict future changes and understand the profound influence of these forces on the planet's ever-changing story.

Conclusion: The Unstoppable Forces of Nature

Volcanoes, earthquakes, and moving tectonic plates are a testament to the ceaseless activity that underpins our planet's dynamic nature. While these processes can bring destruction and upheaval, they also create the breathtaking landscapes and geological wonders that define Earth's beauty and diversity. As we explore the turbulent yet awe-inspiring forces that shape our world, we gain a deeper appreciation for the intricate dance of Earth's geological processes and our place within it.

CHAPTER 14: PAST, PRESENT, AND FUTURE: HOW TIME FLOWS

Time, the enigmatic river that carries us through existence, shapes our experiences, memories, and perceptions. From the distant echoes of the past to the ever-evolving canvas of the present and the infinite possibilities of the future, time is an intangible force that governs our lives and the universe around us. Delving into the profound concept of time unveils its intricate dimensions, the philosophical debates it inspires, and its role in shaping the very fabric of reality.

The Nature of Time: An Elusive Concept

Time is a fundamental aspect of our lives, yet defining it is no simple task. In classical physics, time is often treated as an absolute dimension that flows at a constant rate, independent of our perception. However, modern physics, especially Einstein's theory of relativity, has revealed that time is relative and can be influenced by factors such as gravity and velocity. This revelation challenges our intuitive understanding of time as a fixed and universal construct.

The Arrow of Time: From Past to Future

The arrow of time denotes the apparent one-way progression of events from the past to the present and on to the future. This concept is intimately linked with the second law of thermodynamics, which states that entropy, a measure of disorder, tends to increase over time. This asymmetry in entropy gives rise to our perception of time's directionality, where events become increasingly disordered as time advances.

The Philosophy of Time: A Philosophical Puzzle

Throughout history, philosophers have grappled with the nature of time. Questions about whether time is an objective reality or a human construct, whether the past, present, and future are equally real, and the possibility of time travel have sparked intense debates. Philosophical ideas range from the block universe theory, where the past, present, and future exist simultaneously, to the presentism theory, which asserts that only the present is real.

Einstein's Relativity: A New Paradigm

Albert Einstein's theory of relativity revolutionized our understanding of time. The theory's two components, special relativity and general relativity, demonstrate that time is not constant but can vary depending on factors like speed and gravity. Time dilation, a key consequence of special relativity, implies that time passes slower for an observer in motion relative to a stationary observer. General relativity shows that massive objects warp spacetime, affecting the passage of time in their vicinity.

Time and Space: A Cosmic Tapestry

Einstein's insights also revealed the inseparable relationship between time and space. Spacetime is a four-dimensional continuum that blends the three dimensions of space with the dimension of time. Objects moving through spacetime trace out a path called a worldline, which encapsulates their entire history, from their birth to their eventual fate.

Quantum Mechanics and Time: A Quantum Puzzle

The quantum realm introduces another layer of complexity to the concept of time. Quantum mechanics, the physics that governs the behavior of subatomic particles, challenges our notions of cause and effect. Quantum entanglement, where particles become linked in ways that defy classical logic, raises intriguing questions about the nature of time's flow and the potential for instantaneous communication.

Time in Our Lives: Moments and Memories

Time is a deeply personal experience, influencing the rhythm of our lives and the stories we create. The present moment is a fleeting point of intersection between past and future, where experiences unfold and decisions are made. Memories, often selective and malleable, connect us to the past, shaping our identity and influencing our choices.

The Future: An Open Horizon

The future is a realm of endless possibilities and uncertainties. While we can make educated guesses and plans based on our understanding of the present, the unfolding of events is influenced by countless variables. Quantum uncertainty, chaotic systems, and the intricate interplay of human decisions all contribute to the complexity of predicting the future.

Time Travel: Science Fiction and Reality

The concept of time travel has captured human imagination for centuries, inspiring countless stories and films. While time travel as depicted in science fiction remains speculative, physicists explore theoretical frameworks like wormholes and closed time-like curves that suggest the possibility of traversing time. However, these concepts are fraught with paradoxes and challenges, leaving their realization uncertain.

Conclusion: Navigating the River of Time

Time, like a flowing river, carries us through the landscapes of our lives, weaving together the past, present, and future into a tapestry of existence. As we explore the intricate dimensions of time, we navigate the ever-changing currents that shape our perception of reality and guide our journey through the cosmos. The quest to understand time's essence and its role in shaping the universe is a timeless pursuit that continues to unravel the mysteries of our existence.

CHAPTER 15: SPACE ADVENTURES: STARS, PLANETS, AND BEYOND

The cosmos, a realm of infinite wonder and mystery, beckons humanity to embark on a journey beyond our planet's boundaries. From the dazzling stars that light up the night sky to the enigmatic planets that orbit distant suns, the universe is a tapestry of cosmic marvels waiting to be explored. Embarking on a space adventure allows us to uncover the secrets of the cosmos, from the birth of stars to the potential for extraterrestrial life, expanding our understanding of the vast expanse that surrounds us.

The Universe's Canvas: A Cosmic Symphony

The universe encompasses everything that exists, from the tiniest subatomic particles to the grandest galaxies. It is a vast expanse, billions of years old, that houses a staggering array of celestial objects, each with its own story to tell. Stars, planets, galaxies, black holes, and more compose the universe's cosmic symphony, harmonizing in a dance of gravity, light, and matter.

Stellar Birth and Death: A Cosmic Lifecycle

Stars are born from clouds of gas and dust, where gravitational forces cause these materials to collapse and heat up. Nuclear fusion ignites within their cores, producing the energy that makes stars shine. Depending on their mass, stars can live for millions to billions of years before eventually exhausting their nuclear fuel. Massive stars end their lives in supernova explosions, leaving behind remnants like neutron stars or black holes.

Galaxies: Cosmic Cities of Stars

Galaxies are vast collections of stars, gas, dust, and dark matter held together by gravity. They come in various shapes and sizes, from spiral galaxies like the Milky Way to elliptical and irregular galaxies. Galaxies host diverse environments, including star-forming regions, supermassive black holes at their centers, and cosmic collisions that shape their evolution.

Planetary Exploration: The Final Frontier

Exploring our solar system and beyond has been a monumental endeavor. The solar system consists of the sun, eight planets, their moons, asteroids, and comets. Spacecraft like Voyager, Cassini, and New Horizons have provided invaluable insights into distant worlds, revealing the mysteries of Jupiter's storms, Saturn's rings, and Pluto's icy plains.

Exoplanets: Worlds Beyond Our Solar System

The discovery of exoplanets—planets that orbit stars outside our solar system—has revolutionized our understanding of the cosmos. The Kepler Space Telescope, among others, has identified thousands of exoplanets, some of which reside in the "habitable zone," where conditions might be right for liquid water and potentially life to exist.

Black Holes and Dark Matter: Mysteries of the Cosmos

Black holes, collapsed remnants of massive stars, possess such strong gravity that not even light can escape them. They challenge our understanding of physics and offer insights into the extreme conditions near these cosmic abysses. Dark matter, a mysterious form of matter that doesn't emit light, comprises most of the universe's mass, yet its nature remains one of the greatest unsolved puzzles in astrophysics.

Cosmic Evolution: From the Big Bang to Now

The universe's story begins with the Big Bang—an explosion of energy and matter that initiated its expansion. As the universe expanded and cooled, particles combined to form atoms, which eventually coalesced into galaxies and stars. The universe's evolution, shaped by gravity, cosmic inflation, and dark energy, has led to the complex and diverse cosmos we observe today.

Astrobiology: The Search for Life Beyond Earth

Astrobiology explores the potential for life beyond our planet. The study of extremophiles—organisms that thrive in extreme conditions—offers insights into life's adaptability. Mars, with its history of water and the possibility of subsurface habitats, has been a target for the search for past or present life. The discovery of water on moons like Europa and Enceladus raises intriguing possibilities for life in our solar system.

Interstellar Travel and SETI: Reaching for the Stars

Interstellar travel, a staple of science fiction, remains a distant dream due to the vast distances and technological challenges involved. However, projects like Breakthrough Starshot propose using light-propelled nanocraft to reach nearby star systems. The Search for Extraterrestrial Intelligence (SETI) listens for signals from advanced civilizations, probing the question of whether we are alone in the cosmos.

Conclusion: Our Cosmic Odyssey

Space adventures are humanity's quest to explore the cosmos, to uncover the mysteries that lie beyond our planet's boundaries. From the brilliance of stars to the enigma of black holes, from the exploration of our solar system to the search for life among the stars, our cosmic odyssey opens doors to discovery, wonder, and the pursuit of knowledge. As we gaze into the heavens, we glimpse not only the past and present of the universe but also the potential futures that beckon us to venture further into the grand tapestry of space.

CHAPTER 16: SUPER SMALL WORLD: THE ODDNESS OF QUANTUM PHYSICS

At the heart of the smallest scales of the universe lies a realm of bewildering paradoxes, where particles can be in multiple places at once, information can be transferred instantly across vast distances, and the act of observation alters the very nature of reality. This is the perplexing world of quantum physics, a branch of science that challenges our intuitive understanding of the universe and offers a glimpse into the bizarre nature of the subatomic realm.

The Quantum Revolution: A Paradigm Shift

Quantum physics emerged in the early 20th century, sparked by the works of physicists like Max Planck, Albert Einstein, Niels Bohr, and Erwin Schrödinger. These pioneers shattered classical notions of determinism and offered new insights into the behavior of particles on the smallest scales.

Wave-Particle Duality: A Dual Nature

One of the foundational principles of quantum physics is wave-particle duality. Particles like electrons and photons exhibit both wave-like and particle-like behavior. This duality challenges our classical understanding, suggesting that particles are not strictly confined to a single location and can exhibit interference patterns like waves.

Uncertainty Principle: The Limit of Knowledge

Werner Heisenberg's uncertainty principle states that there are inherent limits to simultaneously knowing a particle's position and momentum with absolute precision. The more accurately we know one property, the less precisely we can know the other. This principle highlights the fundamental randomness and indeterminacy that underlie the quantum realm.

Quantum Entanglement: Spooky Action at a Distance

Quantum entanglement, as famously described by Einstein as "spooky action at a distance," is one of the most mind-boggling aspects of quantum physics. When two particles become entangled, their properties become correlated in such a way that the state of one particle instantaneously influences the state of the other, regardless of the distance separating them.

Superposition: The Power of Possibility

In the quantum world, particles can exist in a superposition of multiple states simultaneously. This means that a particle, such as an electron, can exist in multiple places or states at once until it is observed, collapsing its wave function into a single state. This phenomenon is exemplified in Schrödinger's famous thought experiment involving a cat that is both alive and dead until observed.

Quantum Tunneling: Crossing Impossible Barriers

Quantum tunneling allows particles to pass through energy barriers that classical physics would consider insurmountable. This phenomenon is crucial in explaining how particles like electrons can "jump" through seemingly impenetrable barriers, enabling the operation of many electronic devices and phenomena like nuclear fusion in stars.

The Observer Effect: Influencing Reality

The act of observation in the quantum world has a profound impact on the outcome of experiments. This is known as the observer effect. The presence of an observer can cause particles to behave differently than when they are unobserved, raising questions about the role of consciousness in shaping reality.

Quantum Computing: A New Frontier

Quantum mechanics has given rise to the field of

quantum computing, promising to revolutionize computation by harnessing the peculiar properties of quantum bits (qubits). Quantum computers have the potential to solve complex problems much faster than classical computers, potentially revolutionizing fields like cryptography and optimization.

Interpreting Quantum Reality: Many Worlds and More

The nature of quantum reality has sparked numerous interpretations, each attempting to explain the underlying mechanisms behind quantum phenomena. The Copenhagen interpretation, the many-worlds hypothesis, and the pilot-wave theory are just a few of the proposed ways to make sense of the quantum world, each with its own strengths and challenges.

Conclusion: The Dance of the Subatomic

Quantum physics challenges our intuition, stretches the limits of our imagination, and reminds us that reality is far stranger and more complex than we can fathom. The oddness of quantum physics defies classical notions of causality and determinism, inviting us to explore the mysterious intricacies of the super small world. As we venture into the depths of quantum physics, we are reminded that even in the most puzzling of realms, the pursuit of knowledge continues to illuminate the fascinating dance of the subatomic particles that compose our universe.

CHAPTER 17: HOW LIFE WORKS: INSIDE PLANTS AND ANIMALS

Life, a symphony of complex processes and intricate mechanisms, manifests in the vibrant diversity of plants and animals that populate our planet. From the graceful dance of photosynthesis in plants to the intricate symphony of cells within animals, the inner workings of living organisms are a testament to the marvels of evolution and adaptation. Delving into the biology of plants and animals unveils the fascinating mechanisms that sustain life, drive growth, and enable these organisms to thrive in their respective ecosystems.

Plants: Harnessing the Power of Sunlight

Plants are the Earth's green architects, capturing the energy of the sun and transforming it into chemical energy through photosynthesis. This process involves chloroplasts, specialized organelles within plant cells, capturing light energy and converting it into glucose, the primary source of energy for plants. Oxygen is released as a byproduct, enriching the atmosphere.

Cell Structure: The Building Blocks of Life

Cells are the fundamental units of life in both plants and animals. Plant cells have a rigid cell wall made of cellulose, providing structural support. Animal cells lack cell walls but possess flexible cell membranes. Within cells, organelles like the nucleus, mitochondria, and endoplasmic reticulum play vital roles in processes like DNA replication, energy production, and protein synthesis.

Reproduction and Growth: Creating the Next Generation

Plants and animals reproduce to ensure the survival of their species. Plants use methods like pollination and seed dispersal, while animals employ diverse reproductive strategies, from internal fertilization to the egg-laying behavior of birds and reptiles. Growth is a dynamic process in which cells divide, differentiate, and specialize, leading to the development of tissues, organs, and ultimately, mature organisms.

Circulatory Systems: Pumping Life's Fluid

In animals, circulatory systems transport vital nutrients, oxygen, and waste products throughout the body. Mammals and birds have closed circulatory systems, where blood is confined to vessels, while some invertebrates possess open circulatory systems, where blood flows in spaces called hemocoels. In plants, vascular tissues like xylem and phloem facilitate the transport of water, nutrients, and sugars.

Respiration: Breathing Life In and Out

Respiration is the process of exchanging gases between an organism and its environment. In animals, respiration involves taking in oxygen and expelling carbon dioxide. Mammals and birds have lungs, while fish use gills to extract oxygen from water. Plants engage in photosynthesis during the day and respiration at night, absorbing oxygen and releasing carbon dioxide.

Digestion and Nutrition: Fueling the Machine

Digestion breaks down food into smaller molecules that can be absorbed by cells. Animals have specialized digestive systems tailored to their diets—herbivores, carnivores, and omnivores have distinct adaptations. Plants, unable to move, absorb water and minerals from the soil through their roots, while leaves engage in photosynthesis, producing sugars from carbon dioxide and water.

Nervous and Sensory Systems: Responding to the World

The nervous system enables animals to receive and process information from their environment. In vertebrates, including humans, this system includes the brain, spinal cord, and peripheral nerves. Sensory organs like eyes, ears, and noses allow animals to perceive their surroundings and respond to stimuli, enhancing their chances of survival and reproduction.

Ecosystem Interactions: Finding Balance in Nature

Plants and animals are integral components of ecosystems, interconnected through intricate webs of interactions. Herbivores consume plants, carnivores' prey on other animals, and scavengers clean up the remains. Mutualistic relationships, like pollination and seed dispersal, benefit both plants and animals. These interactions maintain the balance of energy and nutrients within ecosystems.

Adaptation and Evolution: Nature's Experimentation

Life on Earth has evolved through the process of adaptation. Traits that enhance an organism's survival and reproduction are favored by natural selection, leading to the accumulation of advantageous traits in a population over time. This process has given rise to the diversity of life forms we see today, from the smallest microbes to the largest mammals.

Conclusion: A Symphony of Life

The intricacies of life within plants and animals reveal a symphony of biological processes, adaptations, and interactions. From the microscopic dance of cells to the macroscopic splendor of ecosystems, the beauty and complexity of life's inner workings are a testament to the wonders of nature's design. As we continue to unravel the mysteries of life, we gain a deeper appreciation for the exquisite harmony that sustains the myriad forms of life on our planet.

CHAPTER 18: ENERGY EVERYWHERE: WHERE IT COMES FROM AND GOES

Energy, the lifeblood of the universe, propels the dance of particles and the motion of matter. It flows through every facet of existence, from the gentle hum of a bee's wings to the fiery fusion reactions within stars. Understanding the sources of energy and tracing its pathways as it transforms from one form to another unveils the fundamental principles that govern the dynamics of the cosmos and our daily lives.

Forms of Energy: A Kaleidoscope of Possibilities

Energy comes in various forms, each with its own unique properties and roles:

1. **Mechanical Energy**: The energy associated with the motion and position of objects. It includes kinetic energy (energy of motion) and potential energy (energy stored in an object's position).

2. **Thermal Energy**: The energy of particles in a substance due to their motion. Temperature reflects the average thermal energy of particles.

3. **Chemical Energy**: Energy stored in the bonds between atoms and molecules. It's released during chemical reactions, like the combustion of fuels.

4. **Electrical Energy**: Energy carried by moving electrons. It powers our homes, electronics, and industries.

5. **Nuclear Energy**: Energy released during nuclear reactions, such as nuclear fission (splitting of atomic nuclei) and nuclear fusion (combining atomic nuclei).

6. **Radiant Energy**: Energy carried by electromagnetic

waves, including visible light, radio waves, and X-rays.

Energy Sources: Tapping into the Cosmic Reservoir

Energy comes from a variety of sources, each harnessing different natural processes:

1. **Fossil Fuels**: Coal, oil, and natural gas are formed from the remains of ancient plants and animals. Burning these fuels releases stored chemical energy as heat.

2. **Renewable Energy**: Solar, wind, hydroelectric, geothermal, and biomass energy sources harness natural processes that continually replenish themselves.

3. **Nuclear Energy**: Nuclear reactions release tremendous amounts of energy, as seen in nuclear power plants and the sun.

4. **Gravitational Energy**: Gravitational potential energy is harnessed in hydroelectric power plants, where water's potential energy is converted to mechanical energy and then electricity.

Energy Transformations: The Dance of Change

Energy constantly transforms from one form to another, following the laws of thermodynamics:

1. **First Law of Thermodynamics (Law of Conservation of Energy)**: Energy cannot be created or destroyed, only converted from one form to another.

2. **Second Law of Thermodynamics**: The total entropy (disorder) of an isolated system increases over time, suggesting that energy transformations tend to increase disorder.

Energy Efficiency and Loss: The Quest for Conservation

Energy transformations are never 100% efficient; some energy is always lost as heat due to friction, resistance, and other factors. The concept of efficiency measures how much useful energy is produced compared to the total input.

Energy in Daily Life: Powering Our World

Energy is the heartbeat of modern society, powering homes, transportation, industries, and technology. It enables communication, entertainment, and medical advancements.

Global Energy Challenges: Sustainability and Environmental Impact

The world's energy demands are growing, but there's a pressing need to transition to more sustainable and environmentally friendly energy sources. Fossil fuels contribute to air pollution and climate change, while renewable sources offer cleaner alternatives.

Conclusion: The Dance of Energy

Energy courses through the fabric of reality, weaving together the tapestry of existence. It fuels the cosmos, sustains life, and empowers human progress. From the blazing infernos of stars to the quiet hum of a lightbulb, the journey of energy is an eternal dance of transformation, revealing the interconnectedness of all things in the grand symphony of the universe. Understanding this dance is key to our ability to navigate the challenges and opportunities that energy presents in our quest for a sustainable and harmonious future.

CHAPTER 19: HEALING SCIENCE: HOW MEDICINE HELPS US

Medicine, a blend of science, compassion, and innovation, has been humanity's steadfast companion in the journey toward health and well-being. From ancient remedies to cutting-edge technologies, the evolution of medicine reflects our unending quest to understand and heal the human body. Examining the multifaceted world of medicine, from its historical roots to its modern frontiers, reveals the profound impact it has on individuals, communities, and the global landscape.

Historical Foundations: Ancient Wisdom and Traditions

Medicine's roots trace back to ancient civilizations that relied on a blend of empirical observations, folklore, and spiritual beliefs. Ancient Egyptians documented medical knowledge in papyrus scrolls, while ancient Greek physicians like Hippocrates emphasized the importance of observing natural phenomena and using reason to diagnose and treat diseases.

Modern Medicine's Dawn: Scientific Inquiry and Progress

The rise of the scientific method in the Renaissance marked a pivotal moment in medicine's evolution. Andreas Vesalius revolutionized anatomical understanding with his meticulous dissections, while Edward Jenner's smallpox vaccine marked the dawn of immunization. The 19th century saw the discovery of antibiotics, including penicillin by Alexander Fleming, reshaping the landscape of infectious disease treatment.

Medical Specializations: The Mosaic of Expertise

Medicine has blossomed into a tapestry of specialized fields, from cardiology and neurology to oncology and pediatrics. These specialties unite a deep understanding of specific aspects of the human body with cutting-edge diagnostics, treatments, and research.

Diagnostic Tools: Peering into the Body's Secrets

Advancements in diagnostic tools have enabled healthcare professionals to peer inside the body with unprecedented clarity. X-rays, MRIs, CT scans, and PET scans provide intricate images that aid in the diagnosis and monitoring of diseases.

Pharmaceuticals: Unleashing the Power of Molecules

Pharmaceuticals harness the potential of molecules to treat and prevent diseases. From pain relief to life-saving antibiotics, these compounds have extended human life expectancy and improved its quality.

Surgical Innovations: Precision and Progress

Surgical techniques have advanced dramatically, minimizing invasiveness and improving outcomes. Robotic surgery, minimally invasive procedures, and transplantation have transformed the treatment of conditions once deemed untreatable.

Genomics and Personalized Medicine: Tailoring Treatment

The Human Genome Project unlocked the blueprint of our genetic code, paving the way for personalized medicine. Healthcare providers can now tailor treatments to individual genetic variations, enhancing efficacy and reducing side effects.

Global Health and Public Health Initiatives: Tackling Challenges Together

Global health initiatives address health disparities, infectious disease outbreaks, and public health concerns on a global scale. Vaccination campaigns, disease surveillance, and health education are pivotal in improving health outcomes for entire populations.

Ethical Considerations: The Heart of Medicine

Medicine is not solely a science; it's also a field driven by ethics and compassion. Balancing the benefits of treatment with patient autonomy, fostering trust between doctors and patients, and addressing issues like end-of-life care are essential ethical dimensions.

Healthcare Systems: Navigating Complexity

Healthcare systems vary around the world, reflecting social, economic, and cultural contexts. Access to healthcare, insurance, medical infrastructure, and government policies all play critical roles in shaping the health of communities.

Medical Challenges and Frontiers: Never-Ending Exploration

Medicine constantly confronts new challenges, from emerging infectious diseases to chronic conditions and mental health crises. Cutting-edge research explores stem cell therapies, regenerative medicine, and artificial intelligence's role in diagnosis and treatment.

Conclusion: The Healing Art

Medicine stands as both a science and an art—a pursuit driven by the marriage of knowledge and empathy. It's a beacon of hope that reaches across time and cultures, offering relief from suffering, extending life, and illuminating the depths of human resilience. As medicine continues to evolve, it's a reminder that the quest to heal is a journey that unites humanity in its shared pursuit of health, vitality, and a better future.

CHAPTER 20: FOOD: FROM FARM TO FORK, AND WHY IT MATTERS

Food is more than sustenance; it's a cornerstone of culture, a source of nourishment, and a reflection of our interconnectedness with the natural world. The journey of food, from its origins on farms to the tables where we gather, is a complex and vital process that shapes our health, economies, and environment. Understanding the intricacies of this journey, and the reasons why it matters, unveils the profound impact food has on our lives and the global landscape.

Farming and Agriculture: Cultivating the Earth's Bounty

Farming, the cornerstone of food production, encompasses a myriad of practices that transform soil, water, and sunlight into a harvest of crops and livestock. Traditional methods have evolved alongside modern technologies, giving rise to conventional and sustainable farming practices.

Food Supply Chains: A Complex Web of Distribution

The journey from farm to fork involves intricate supply chains that connect producers, processors, distributors, retailers, and consumers. These chains ensure that food reaches its intended destinations efficiently, although the complexity can sometimes lead to issues like food waste and unequal access to nutritious options.

Global Food Trade: Navigating a Global Marketplace

Food trade transcends borders, allowing regions to share their agricultural strengths and meet diverse consumer demands. While it fosters international collaboration, it also exposes vulnerabilities, such as disruptions due to geopolitical tensions, climate events, or supply chain interruptions.

Nutrition and Health: Fueling Our Bodies and Minds

Food is the fuel that powers our bodies and minds, impacting our physical well-being and cognitive function. A balanced diet rich in nutrients supports growth, immunity, and overall health. However, access to nutritious foods is not universal, leading to health disparities and diet-related diseases.

Cultural Significance: The Tapestry of Tradition

Food is intertwined with cultural identity, carrying the flavors, traditions, and stories of generations. It unites families and communities in shared experiences, celebrations, and rituals. Cultural diversity enriches the culinary landscape, fostering appreciation and understanding.

Environmental Impact: Treading Lightly on Earth

Food production has significant environmental implications. Agriculture can contribute to deforestation, water pollution, and greenhouse gas emissions. Sustainable practices, like organic farming, regenerative agriculture, and reducing food waste, are pivotal in mitigating these impacts.

Food Security: A Global Challenge

Food security, ensuring all individuals have access to sufficient, safe, and nutritious food, is a global challenge. Economic disparities, climate change, and conflicts can threaten food availability, leading to hunger and malnutrition in vulnerable populations.

Technology and Innovation: Navigating the Future

Technological advancements, from genetically modified organisms (GMOs) to precision agriculture and lab-grown meat, are reshaping the food landscape. While these innovations offer opportunities for increased productivity and reduced environmental impact, they also raise ethical, safety, and regulatory questions.

Consumer Choices: Shaping the Food System

Consumers play a pivotal role in shaping the food system through their choices. Preferences for organic, locally sourced, or ethically produced foods drive market trends, influencing how food is grown, processed, and distributed.

Food Waste: A Global Conundrum

Food waste, a staggering issue, occurs at various points in the food journey. From farm surplus to consumer leftovers, food waste has economic, environmental, and ethical implications, urging efforts to reduce waste and increase efficiency.

Food and Sustainability: Balancing Needs

Achieving a sustainable food system involves balancing ecological, economic, and social factors. This includes addressing issues like soil health, water conservation, biodiversity preservation, fair labor practices, and access to nutritious food for all.

Conclusion: A Plate of Possibilities

Food is more than a daily necessity; it's a gateway to understanding the complexities of our world. From the farmer's toil to the chef's creativity, from cultural heritage to global interconnectedness, food encapsulates a story of abundance and scarcity, tradition and innovation. As we savor each bite and consider the journey that brought it to our table, we recognize the transformative power of food—a power that extends beyond our plates to influence the health of our planet and the well-being of generations to come.

CHAPTER 21: STUFF AROUND US: MATERIALS AND WHAT THEY DO

The world around us is a rich tapestry of materials, each with its own unique properties and purposes. From the clothes we wear to the buildings we inhabit, the materials we use shape our lives, industries, and societies. Delving into the diverse realm of materials science reveals the fascinating nature of matter, the ways in which materials are manipulated, and the profound impact they have on our daily experiences.

Matter and Its States: Building Blocks of the Universe

Matter, the stuff of the universe, exists in different states: solid, liquid, gas, and plasma. The arrangement and motion of atoms and molecules determine these states and their properties. Understanding these fundamental states helps us grasp the behavior of materials under various conditions.

Elements and Compounds: Foundations of Matter

Elements are the purest forms of matter, each consisting of atoms with a distinct number of protons. Compounds are combinations of elements in specific ratios, creating molecules with unique properties. The periodic table categorizes elements by their properties and relationships.

Properties of Materials: Unveiling their Nature

Materials possess a spectrum of properties, including physical, mechanical, thermal, and electrical characteristics. These properties influence how materials are used in different applications. For example, the hardness of a material determines its suitability for tools or construction.

Metals, Polymers, and Ceramics: A Diverse Array

Materials can be classified into three broad categories: metals, polymers, and ceramics.

1. **Metals**: Known for their strength and conductivity, metals are used in infrastructure, electronics, and transportation due to their versatility and ability to conduct heat and electricity.

2. **Polymers**: Polymers are large molecules made up of repeating units. Plastics, a subset of polymers, are found in countless products, from packaging to medical devices, due to their lightweight and moldable nature.

3. **Ceramics**: Ceramics are non-metallic, inorganic materials that are often heat-resistant and have diverse applications, including pottery, electronics, and aerospace components.

Composites: Combining Strengths

Composites blend different materials to capitalize on their individual strengths. Reinforced concrete, fiberglass, and carbon fiber-reinforced polymers are examples of composites that offer enhanced properties by combining materials with complementary characteristics.

Materials Processing: Shaping the Future

Materials processing transforms raw materials into usable forms through techniques like casting, forging, machining, and 3D printing. Advances in processing methods have revolutionized industries, enabling intricate designs, reduced waste, and enhanced performance.

Nanomaterials: A World of the Minuscule

Nanotechnology explores materials at the nanoscale, where properties can differ dramatically from those on a larger scale. Nanomaterials are used in electronics, medicine, and environmental applications, promising revolutionary breakthroughs.

Sustainable Materials: Minimizing Impact

As environmental concerns mount, sustainable materials gain prominence. Biomaterials, recycled materials, and environmentally friendly alternatives reduce the ecological footprint of products and systems.

Materials in Engineering: Pushing Boundaries

Materials science is integral to engineering innovations. High-performance materials enable the construction of skyscrapers, the development of advanced medical devices, and the creation of efficient transportation systems.

Cultural Significance: The Art of Materials

Materials are embedded in culture and history, shaping artistic expression and societal identity. Architecture, fashion, and art showcase the creative potential of materials, influencing aesthetics and innovation.

Future Challenges: Balancing Progress and Sustainability

Materials science faces challenges related to resource scarcity, waste management, and ethical considerations. The quest for new materials that are both functional and environmentally responsible is ongoing.

Conclusion: The Essence of Our World

Materials are the building blocks of our world, shaping the objects we interact with and the environments we inhabit. The quest to understand, manipulate, and harness materials reflects humanity's inherent curiosity and creativity. As we continue to explore the endless possibilities of matter, we shape a future where materials serve not only our needs but also the well-being of the planet and generations to come.

CHAPTER 22: OUR AMAZING BRAIN: HOW IT CONTROLS US

The human brain, a marvel of complexity and ingenuity, sits at the helm of our existence, orchestrating the symphony of thoughts, emotions, memories, and actions that define who we are. Its intricate network of neurons and synapses empowers us to navigate the world, learn from experiences, and adapt to changing circumstances. Understanding the inner workings of this remarkable organ unveils the mysteries of consciousness, cognition, and the delicate balance that shapes our lives.

Anatomy and Structure: The Architect of Thought

The brain is divided into distinct regions, each responsible for specific functions. The cerebral cortex, the outermost layer, plays a central role in conscious thought, language, memory, and decision-making. Deep within the brain, structures like the hippocampus and amygdala regulate memory and emotions, while the brainstem controls vital functions like breathing and heart rate.

Neurons: Building Blocks of Communication

Neurons are the brain's messengers, communicating through electrical impulses and chemical signals. Billions of neurons connect to form intricate neural pathways, enabling the brain to process information, make decisions, and generate consciousness.

Synapses: The Nexus of Connection

Synapses are the tiny gaps between neurons where information is transmitted. When a neuron fires an electrical impulse, it releases neurotransmitters that cross the synapse, conveying signals to the next neuron. This complex web of connections allows thoughts and behaviors to emerge.

Neuroplasticity: Shaping and Reshaping the Brain

Neuroplasticity is the brain's ability to adapt and rewire itself based on experiences. It underlies learning, memory, and recovery from injury. Throughout life, new connections form and unused ones weaken, sculpting the brain's structure and function.

Consciousness and Perception: The Theater of Experience

The brain constructs our perception of reality. Sensory information from the environment is processed, integrated, and interpreted, forming our conscious experience of sights, sounds, tastes, and more.

Emotions and Memories: The Colors of Experience

Emotions and memories are intricately linked to brain activity. The amygdala processes emotions, while the hippocampus is crucial for forming new memories. The interplay of these regions contributes to our emotional experiences and the recollection of past events.

Language and Communication: A Symphony of Words

Language processing involves multiple brain areas, including Broca's area for speech production and Wernicke's area for comprehension. The brain's ability to process and generate language is fundamental to human communication and expression.

Learning and Intelligence: The Quest for Knowledge

Learning is the brain's mechanism for acquiring new information and skills. Intelligence, influenced by genetic factors and environmental experiences, is the brain's ability to reason, solve problems, and adapt to challenges.

Motor Control: Dance of Precision

Motor control involves a complex interplay between the brain's motor cortex, basal ganglia, and cerebellum. These regions coordinate movements, from the simplest tasks like grasping objects to the intricate artistry of dance.

Disorders and Conditions: The Fragility of Balance

Brain disorders, such as Alzheimer's disease, Parkinson's disease, and depression, underscore the delicate balance required for optimal brain function. Research seeks to unravel the mechanisms behind these conditions and develop effective treatments.

Ethical Considerations: The Frontier of Neuroscience

Advancements in brain research raise ethical questions about privacy, consent, and the potential for cognitive enhancement. Brain-computer interfaces, neuroimaging, and cognitive enhancement technologies challenge our understanding of consciousness and autonomy.

The Future of Brain Science: Unveiling the Mind's Mysteries

Brain science is on the cusp of breakthroughs that could revolutionize our understanding of consciousness, brain disorders, and artificial intelligence. Mapping the brain's connectome, simulating neural networks, and decoding thoughts are all exciting frontiers.

Conclusion: The Essence of Identity

The human brain stands as a testament to the marvels of evolution and the complexities of consciousness. It's the epicenter of our thoughts, emotions, and actions—a reflection of the intricate interplay between biology, environment, and experience. The journey to understand this extraordinary organ is a voyage into the heart of our identity, unveiling the depths of human potential and the uncharted territories of the mind.

CHAPTER 23: GENES AND TRAITS: WHAT MAKES YOU, YOU

Deep within the intricate code of our DNA lies the blueprint of our existence, the keys to our physical features, behaviors, and susceptibilities. The relationship between genes and traits is a captivating journey into the realms of genetics, heredity, and the fascinating interplay between our genetic makeup and the environment that shapes us. Unraveling the mysteries of genes and traits provides insights into our uniqueness as individuals and the threads that connect us to our ancestors and future generations.

Genes and DNA: The Foundations of Life

Genes are segments of DNA that contain instructions for building proteins, the molecular workhorses of our bodies. DNA, the double-stranded helix, carries the genetic information that encodes our traits. These traits encompass physical attributes, such as eye color and height, as well as physiological processes and predispositions to certain conditions.

Inheritance and Genetics: The Heritage of Traits

Heredity is the passage of genetic information from one generation to the next. The principles of inheritance were elucidated by Gregor Mendel, who discovered dominant and recessive traits through his pea plant experiments. Genetic traits are inherited from our parents, with each parent contributing one set of genes to form our unique genetic makeup.

Genotype and Phenotype: The Genetic Identity

Genotype refers to the specific genetic composition of an individual, including the alleles (gene variants) they carry. Phenotype encompasses the observable traits resulting from the interaction between an individual's genes and their environment.

Mendelian and Non-Mendelian Inheritance: Patterns of Diversity

While Mendelian genetics describe straightforward inheritance patterns, non-Mendelian inheritance involves more complex interactions between genes. Codominance, incomplete dominance, and polygenic inheritance contribute to the diversity of traits observed in populations.

Genetic Variation: Embracing Diversity

Genetic variation is the cornerstone of evolution. Mutations, changes in the DNA sequence, introduce new alleles into populations. This diversity provides the raw material for natural selection to shape traits that increase an organism's fitness.

Nature vs. Nurture: A Dynamic Interaction

The nature vs. nurture debate explores the interplay between genetic predisposition and environmental influences. While genes provide the foundation, environmental factors such as nutrition, upbringing, and exposure to toxins also shape our traits.

Genomics: Decoding the Genetic Library

Genomics involves studying the entirety of an organism's genetic material. Advances in genomics have led to breakthroughs in understanding diseases, identifying genetic markers, and tailoring personalized medical treatments.

Epigenetics: Beyond the DNA Sequence

Epigenetics refers to modifications to DNA that affect gene expression without altering the underlying genetic code. These modifications are influenced by environmental factors and can be passed on to subsequent generations.

Genetic Disorders: From Mutations to Consequences

Genetic disorders result from mutations that disrupt the normal functioning of genes. These disorders can be inherited or arise spontaneously. Some, like cystic fibrosis and sickle cell anemia, are caused by mutations in a single gene, while others involve multiple genes or complex interactions.

Genetic Engineering and Ethics: Manipulating Nature

Advancements in genetic engineering raise ethical dilemmas regarding the manipulation of genes for therapeutic or enhancement purposes. CRISPR-Cas9 technology, for instance, allows precise gene editing, opening possibilities for treating genetic disorders but also sparking debates about its implications.

Human Identity and Genetic Ancestry: Uncovering Roots

Genetic testing and ancestry tracing provide insights into our origins and connections to diverse populations across time and space. They also underscore the shared heritage that unites humanity.

Conclusion: The Mosaic of Traits

Genes and traits weave a tapestry of human diversity and individuality. They reflect the intricate dance between our inherited genetic makeup and the dynamic environment that sculpts our lives. The journey to understand this relationship is a testament to our pursuit of knowledge, our reverence for life's complexity, and our quest to uncover the intricate threads that make each of us a unique expression of the human story.

CHAPTER 24: SPACE MOVES: PLANETS AND ORBITS IN ACTION

The celestial dance of planets and their orbits is a mesmerizing display of the fundamental laws that govern our universe. From the graceful orbits of Earth and its neighbors to the dynamic forces that shape these paths, the study of planetary motion unveils the profound elegance of celestial mechanics and our place within the cosmic expanse. Delve into the intricacies of planetary orbits and the gravitational forces that propel these celestial bodies on their cosmic journeys.

Kepler's Laws: Unveiling Planetary Orbits

Johannes Kepler's laws of planetary motion laid the foundation for understanding how planets move in space:

1. **Law of Orbits**: Planets move in elliptical orbits, with the Sun at one of the two foci.

2. **Law of Equal Areas**: A line connecting a planet to the Sun sweeps out equal areas in equal intervals of time. This implies that a planet moves faster when closer to the Sun (perihelion) and slower when farther away (aphelion).

3. **Law of Harmonies**: The square of the orbital period of a planet is directly proportional to the cube of the semi-major axis of its orbit. This law relates the time it takes for a planet to orbit the Sun to its distance from the Sun.

Orbital Elements: The Blueprint of Motion

To describe planetary orbits, astronomers use orbital elements. These include the semi-major axis (average distance from the Sun), eccentricity (how elongated the orbit is), inclination (angle between the orbit and a reference plane), and more. These elements determine the unique path each planet follows.

Gravitational Forces: The Cosmic Tug

The force that keeps planets in their orbits is gravity. Isaac Newton's law of universal gravitation states that every mass attracts every other mass with a force proportional to the product of their masses and inversely proportional to the square of the distance between them.

Elliptical Orbits: Paths of Precision

The elliptical shape of planetary orbits ensures the conservation of angular momentum as a planet moves closer to the Sun (increasing its speed) and farther away (decreasing its speed). This ensures that planets obey the laws of Kepler while experiencing gravitational forces.

Tidal Forces: Cosmic Sculptors

Gravitational forces from celestial bodies can create tidal forces on one another. These forces can stretch and deform planets and moons, leading to phenomena like tides on Earth and tidal heating on moons like Io around Jupiter.

Planetary Resonances: Dancing to the Beat of the Cosmos

Planets can interact with each other gravitationally, leading to resonances where their orbital periods are related by simple integer ratios. The most famous example is Pluto and Neptune, which are in a 2:3 resonance, meaning that for every two orbits Pluto completes, Neptune completes three.

Orbital Mechanics and Space Exploration: Navigating the Cosmos

Understanding orbital mechanics is essential for space exploration. Launch windows, interplanetary trajectories, and rendezvous maneuvers all depend on precise calculations of orbital mechanics.

Exoplanet Orbits: Diverse Worlds Beyond

The study of exoplanets, planets orbiting stars outside our solar system, reveals a diverse array of orbits and planetary systems. Some exoplanets have eccentric orbits, while others orbit close to their host stars in what's called the "habitable zone."

The Dance Continues: An Ever-Evolving Cosmos

The motion of planets and the celestial ballet of orbits continue to captivate astronomers and inspire new discoveries. Gravitational interactions, perturbations, and the dynamism of the universe ensure that planetary orbits are not static but part of a dynamic, ever-changing cosmic choreography.

Conclusion: Harmonies in the Heavens

Planetary orbits are a manifestation of the grand symphony of celestial mechanics. From the precise laws of Kepler to the gravitational dance choreographed by Newton's insights, the motions of planets around stars remind us of the harmony that underlies the cosmos. As we gaze at the night sky, we witness the eternal rhythms of space moves, a testament to the beauty and elegance that shape the universe.

CHAPTER 25: WAVES ARE EVERYWHERE: LIGHT, SOUND, AND MORE

Waves, the rhythmic undulations that propagate through various mediums, are omnipresent in the fabric of reality. They form the backbone of communication, the essence of music, and the foundation of our understanding of the universe. From the mesmerizing dance of light to the symphony of sound, the study of waves unveils the interconnected nature of physical phenomena and the profound ways in which they shape our perception of the world.

Nature of Waves: Oscillations in Motion

Waves are characterized by their ability to transport energy without transporting matter. They arise from oscillations, the back-and-forth movements of particles or fields within a medium. Waves can be classified into two broad categories: mechanical waves, which require a medium to travel through, and electromagnetic waves, which can propagate through a vacuum.

Mechanical Waves: Transferring Energy Through Matter

Mechanical waves involve the displacement of particles within a medium. Examples include:

1. **Sound Waves**: Vibrations in air molecules create sound waves that travel as compressions and rarefactions. The frequency of these waves determines the pitch we perceive.

2. **Water Waves**: Disturbances in water's surface generate waves that can vary from gentle ripples to powerful tsunamis.

3. **Seismic Waves**: Earthquakes create seismic waves

that propagate through the Earth's interior, providing insights into its structure.

Electromagnetic Waves: The Dance of Fields

Electromagnetic waves, which include light, are composed of oscillating electric and magnetic fields that can travel through a vacuum. They encompass a broad spectrum, from radio waves to gamma rays, each with unique properties and applications.

Wave Characteristics: Wavelength, Frequency, and More

Waves are defined by key characteristics:

1. **Wavelength**: The distance between successive peaks or troughs of a wave. Longer wavelengths correspond to lower frequencies.

2. **Frequency**: The number of complete oscillations of a wave that pass a given point per unit time. It's inversely related to wavelength.

3. **Amplitude**: The height of a wave's peak or depth of its trough. It reflects the energy carried by the wave.

Wave Interactions: The Dance of Overlaps

When waves encounter each other, they interact in various ways:

1. **Superposition**: When waves overlap, their displacements combine to form a resultant waveform.

2. **Interference**: Interference occurs when two waves overlap and their displacements add or cancel each other out, resulting in constructive or destructive interference.

3. **Diffraction**: Waves can bend around obstacles or spread out as they pass through openings. This phenomenon is called diffraction.

Wave Phenomena: From Doppler Effect to Polarization

Waves exhibit intriguing phenomena that affect our perception:

1. **Doppler Effect**: The apparent change in frequency of a

wave due to the relative motion between the source and the observer. It's responsible for the shift in pitch of a siren as it approaches or recedes.

2. **Polarization**: Certain waves, like light, exhibit polarization, where their oscillations are confined to a specific plane.

Wave Applications: From Communication to Medicine

Waves have a multitude of practical applications:

1. **Communication**: Radio waves, microwaves, and optical fibers transmit information over long distances through modulation techniques.

2. **Medical Imaging**: Ultrasound and X-rays are used for medical imaging, providing non-invasive insights into the body's interior.

3. **Spectroscopy**: The study of how different substances interact with electromagnetic waves helps identify materials and analyze their composition.

Wave-Particle Duality: A Quantum Conundrum

In quantum mechanics, particles like electrons and photons exhibit both particle and wave-like behaviors, a phenomenon known as wave-particle duality. This duality challenges our classical understanding of waves and particles.

Conclusion: The Symphony of Nature

Waves are the threads that weave the tapestry of our universe, connecting the realms of light, sound, and more. They reveal the interconnectedness of physical phenomena, from the vibrations that create music to the invisible dance of electromagnetic fields. Our understanding of waves not only enhances our scientific knowledge but also deepens our appreciation for the harmonious rhythms that govern the natural world, inviting us to explore the rich symphony of waves that surround and shape us.

CHAPTER 26: TINY HIDDEN WORLDS: WHAT MICROSCOPES SHOW US

The world of the microscopic, a realm hidden from our naked eyes, is unveiled through the remarkable invention of microscopes. These intricate devices have revolutionized our understanding of life, matter, and the intricate structures that make up our universe. From the dazzling intricacies of cells to the hidden wonders of atoms, the study of these tiny hidden worlds offers a gateway to exploring the beauty and complexity that exists beyond our ordinary perception.

The Birth of Microscopy: Revealing the Unseen

The invention of the microscope in the 17th century by Antonie van Leeuwenhoek opened a new frontier of exploration. With lenses that magnified objects many times over, microscopes provided a glimpse into the minuscule wonders that had long remained invisible.

Types of Microscopes: Peering into the Minuscule

Several types of microscopes have been developed to suit different purposes:

1. **Light Microscopes**: Using visible light to illuminate specimens, these microscopes reveal structures down to the cellular level.

2. **Electron Microscopes**: With much higher magnification, electron microscopes use beams of electrons to create detailed images, allowing us to observe molecules and even individual atoms.

Cellular Mysteries: Life at the Microscopic Level

Microscopes have revealed the intricate world of cells, the

fundamental units of life. From bacteria to complex eukaryotic cells, these structures house organelles, genetic material, and the complex processes that sustain life.

Microbes Unveiled: Tiny Organisms with a Huge Impact

Microscopes have given us insight into the realm of microbes, the invisible organisms that shape our world. Bacteria, viruses, and other microorganisms play pivotal roles in health, disease, and the environment.

Molecular Marvels: Peering into the Atomic Level

Advancements in microscopy have allowed us to observe molecular structures, including proteins, DNA, and other biomolecules. Techniques like X-ray crystallography and cryo-electron microscopy have transformed our understanding of the building blocks of life.

Materials at the Nanoscale: Engineering New Frontiers

Nanoscience and nanotechnology explore the world of materials and structures at the nanometer scale. Scanning probe microscopes and atomic force microscopes enable scientists to manipulate and study individual atoms, opening doors to innovative materials and applications.

Medical Insights: Diagnosing and Treating Disease

Microscopy plays a crucial role in medicine. Pathologists use microscopes to examine tissue samples, diagnose diseases, and guide treatment decisions. Advances in imaging techniques have also revolutionized non-invasive medical imaging.

Environmental Impact: Studying the Microscopic Ecosystem

Microscopes are essential tools for studying environmental samples. Microbial communities, soil composition, and pollution levels are all investigated using microscopic techniques.

Art and Aesthetics: The Microscopic Beauty

Microscopic images often reveal stunning patterns, colors, and textures that transcend scientific study. Art and science intersect as these images inspire awe and appreciation for the hidden beauty of the microcosmos.

Technological Challenges: Pushing the Limits

The advancement of microscopy techniques faces technological challenges. Improving resolution, developing non-invasive imaging methods, and handling delicate samples are ongoing areas of research.

Conclusion: Unveiling the Invisible

Microscopes have allowed us to transcend the limits of our senses, revealing hidden worlds of astonishing complexity and beauty. From the tiniest cells to the intricate structures of matter, they have provided us with a window into the universe that would otherwise remain unseen. The study of these tiny hidden worlds not only enriches our scientific knowledge but also fuels our curiosity and wonder, reminding us that there is much more to the universe than meets the eye.

CHAPTER 27: OCEAN MYSTERIES: EXPLORING THE SEAS

The oceans, Earth's vast and enigmatic blue expanses, have captivated human curiosity for centuries. Covering more than 70% of the planet's surface, the oceans are teeming with life, mysteries, and untold wonders that continue to intrigue scientists, adventurers, and explorers alike. From the depths of the abyss to the intricate ecosystems of coral reefs, the study of oceanography unveils the awe-inspiring complexities and hidden treasures that lie beneath the waves.

The Depths Unveiled: A Journey into the Abyss

The ocean is divided into various zones, each with unique characteristics and life forms:

1. **Epipelagic Zone**: The sunlit surface layer where most marine life thrives through photosynthesis.

2. **Mesopelagic Zone**: Known as the twilight zone, it lies beneath the sunlight penetration and houses creatures adapted to low light.

3. **Bathypelagic Zone**: Deeper still, this zone is home to a host of bizarre and bioluminescent organisms.

4. **Abyssopelagic Zone**: The deep abyss, characterized by extreme pressures and minimal light, is inhabited by some of the strangest creatures on Earth.

5. **Hadalpelagic Zone**: The deepest parts of the ocean, including the Mariana Trench, harbor mysterious ecosystems adapted to extreme conditions.

Marine Biodiversity: A Universe Beneath the Waves

The oceans support an astonishing array of life, from the smallest plankton to the largest whales. Coral reefs are among the most diverse ecosystems, providing habitats for countless species and playing a crucial role in maintaining marine biodiversity.

Ocean Currents: Earth's Circulatory System

Ocean currents, driven by wind, temperature, and salinity gradients, circulate water around the globe. These currents influence climate, weather patterns, and the distribution of marine life.

Climate Regulation: The Ocean's Role as a Climate Stabilizer

The oceans act as a massive heat sink, regulating Earth's climate by absorbing and releasing heat over long periods. They also play a vital role in the carbon cycle, absorbing and storing vast amounts of carbon dioxide.

Ocean Exploration: Unveiling the Unknown

Exploring the oceans has been a journey of discovery and technological innovation. Submersibles, remotely operated vehicles (ROVs), and autonomous underwater vehicles (AUVs) allow scientists to study the deep sea and its inhabitants.

Deep-Sea Mysteries: Venturing into the Abyss

The deep sea holds mysteries yet to be unraveled. Hydrothermal vents, where superheated water spews from the seafloor, support unique ecosystems that thrive in complete darkness.

Marine Conservation: Preserving the Precious Ecosystems

Human activities, such as overfishing, pollution, and climate change, pose significant threats to marine ecosystems. Marine protected areas, sustainable fishing practices, and efforts to reduce plastic waste are crucial for safeguarding the oceans.

Oceanographic Research: Unraveling the Complexities

Oceanography encompasses a wide range of disciplines, from physical oceanography (studying currents and water properties) to biological oceanography (exploring marine life and ecosystems). These disciplines collaborate to form a holistic understanding of the oceans.

Future Challenges: Balancing Exploration and Conservation

The future of ocean exploration requires a delicate balance between satisfying human curiosity and protecting the fragile marine environment. Emerging technologies, sustainable practices, and international cooperation will shape the way forward.

Conclusion: The Infinite Blue Horizon

The oceans, shrouded in their depths and mysteries, hold the promise of countless revelations. From the extraordinary creatures dwelling in the abyss to the intricate dance of ocean currents, the seas are a testament to the beauty and complexity of Earth's interconnected systems. As we embark on journeys to explore the oceans' secrets, we must remember the importance of stewardship, preserving these precious ecosystems for generations to come. The oceans remain a boundless frontier, inviting us to unravel their mysteries, appreciate their wonders, and learn from their wisdom as we navigate the tides of discovery.

CHAPTER 28: GREEN WORLD: HOW PLANTS GROW AND LIVE

Plants, the silent architects of Earth's ecosystems, shape the very foundation of life as we know it. From the towering trees of lush forests to the delicate wildflowers in open meadows, the world of plants encompasses a fascinating spectrum of forms, functions, and interactions. Exploring the intricacies of how plants grow, adapt, and coexist within their environments unveils the astonishing complexities that drive the vitality and balance of our green world.

Plant Anatomy: The Structures of Life

Plants are composed of various structures, each serving a specific function:

1. **Roots**: Anchoring plants in the soil, roots also absorb water and nutrients vital for growth.

2. **Stems**: Stems provide support for leaves and flowers, as well as transport systems for water, nutrients, and food.

3. **Leaves**: Leaves are the primary sites of photosynthesis, where plants convert sunlight into energy.

4. **Flowers**: Flowers play a crucial role in reproduction, attracting pollinators and producing seeds.

Photosynthesis: The Miracle of Green Alchemy

Photosynthesis is the process by which plants capture sunlight, carbon dioxide, and water to produce energy-rich glucose and oxygen. This intricate biochemical dance sustains life on Earth, releasing oxygen into the atmosphere and providing the foundation of the food chain.

Plant Growth and Development: The Journey to Maturity

Plants exhibit a lifecycle that includes seed germination, growth, flowering, pollination, and seed production. The complex interplay of hormones, environmental cues, and genetic factors orchestrates these stages.

Adaptations: Thriving in Diverse Habitats

Plants have evolved a myriad of adaptations to survive in various environments:

1. **Xerophytes**: Desert plants, like cacti, have adapted to arid conditions by reducing water loss through modified leaves and stems.

2. **Hydrophytes**: Aquatic plants have evolved features like air sacs and submerged leaves to thrive in water.

3. **Epiphytes**: These plants, like orchids, grow on other plants to access sunlight and nutrients.

Plant-Animal Interactions: The Web of Life

Plants interact with animals in various ways:

1. **Pollination**: Insects, birds, and other animals assist in pollination, ensuring the continuation of plant species.

2. **Seed Dispersal**: Animals help disperse seeds by eating fruits and transporting seeds to new locations.

3. **Symbiosis**: Mycorrhizal fungi form mutually beneficial relationships with plants, aiding in nutrient uptake.

Human and Environmental Impact: The Power of Plants

Plants play a vital role in mitigating climate change, acting as carbon sinks that absorb CO_2 from the atmosphere. They also provide habitat for countless species and contribute to soil fertility and erosion control.

Agriculture and Food Security: Feeding the World

The cultivation of crops has been central to human civilization. Advances in agricultural science have led to increased yields and improved crop varieties, addressing global food security challenges.

Conservation and Biodiversity: Protecting Plant Diversity

The loss of plant species threatens biodiversity and ecosystems. Efforts to conserve plant diversity include seed banks, habitat preservation, and restoration projects.

Plant Research and Innovation: Exploring the Frontiers

Botanists and researchers are unlocking the secrets of plants at the molecular level, uncovering mechanisms for stress tolerance, disease resistance, and more. These discoveries have implications for agriculture, medicine, and environmental conservation.

Conclusion: Guardians of Life's Balance

Plants are the unsung heroes that sustain life on Earth. From the grandest trees to the tiniest mosses, they are the threads that weave the intricate tapestry of ecosystems, providing oxygen, nourishment, and habitat for a multitude of organisms. Understanding the complexities of plant growth, adaptation, and interaction underscores the delicate harmony that exists within the green world—a harmony that deserves our respect, protection, and continued exploration as we navigate the path toward a sustainable and balanced future.

CHAPTER 29: OUR EYES PLAY TRICKS: SEEING ISN'T ALWAYS BELIEVING

The human visual system is a marvel of evolution, allowing us to perceive and navigate the world around us. Yet, this intricate system is not without its quirks and limitations. From optical illusions that challenge our perception to the brain's role in shaping our reality, the topic of how our eyes play tricks reveals the complex interplay between biology, psychology, and the way we experience the world.

Vision: A Symphony of Light and Perception

The process of vision involves a delicate choreography of light, optics, and neural processing:

1. **Light**: Photons of light enter the eye and pass through the cornea and lens, focusing on the retina—a layer of light-sensitive cells.

2. **Retina**: Photoreceptor cells, rods for low-light vision and cones for color and detail, convert light into electrical signals.

3. **Optic Nerve**: These signals travel along the optic nerve to the brain's visual cortex, where they are processed into the images we perceive.

Perception and Illusions: The Brain's Interpretive Dance

Optical illusions demonstrate the brain's role in interpreting visual information. These illusions exploit the brain's tendency to fill in gaps, make assumptions, and prioritize certain cues over others.

1. **Müller-Lyer Illusion**: Lines of equal length appear unequal due to the influence of arrows pointing inward

or outward.

2. **Hermann Grid Illusion**: Spots appear at intersections of a grid due to contrast enhancement and lateral inhibition in the visual system.

3. **Color Afterimages**: Staring at a colored image then shifting gaze to a neutral background creates an afterimage in complementary colors, as the brain's color receptors adapt.

Cognitive Biases and Perceptual Errors: Mind's Influence on Sight

Our brains process visual information based on prior experiences and cognitive biases:

1. **Confirmation Bias**: We tend to see what we expect or want to see, filtering out contradictory information.

2. **Cognitive Dissonance**: Our brains might alter our perception to align with our beliefs, creating an illusion of consistency.

Depth Perception: The Art of Seeing in 3D

Depth perception allows us to gauge distances and perceive the world in three dimensions. The brain uses cues like binocular disparity (the slight difference between images seen by each eye) and monocular cues (perspective, shading, overlap) to construct depth.

Visual Attention: The Brain's Spotlight

Visual attention is the brain's way of focusing on specific aspects of a scene while filtering out distractions. This process influences our perception and memory formation.

Synesthesia: The Blurring of Senses

Synesthesia is a phenomenon where sensory experiences blend together. Some people may see colors when they hear music or associate tastes with certain words. This blending of senses challenges our traditional understanding of perception.

Virtual Reality and Perception: Manipulating Reality

Virtual reality (VR) technology alters our visual perception by immersing us in computer-generated environments. It demonstrates how our brains can be tricked into perceiving unreal situations as real.

The Unconscious Mind: Beneath the Surface of Awareness

The brain often processes visual information unconsciously. Studies have shown that our brains recognize complex patterns and make decisions before we are consciously aware of them.

Conclusion: The Complex Dance of Perception

Our eyes capture only a fraction of the electromagnetic spectrum, and our brains further interpret and mold the information we perceive. The relationship between vision, cognition, and reality is intricate and multifaceted. The ways in which our eyes play tricks on us remind us that our perception is a blend of sensory input, past experiences, and cognitive processes—a dance of complexity that challenges the very nature of reality and highlights the intricate workings of the human mind.

CHAPTER 30: HEAT AND ENERGY: MAKING THINGS GO

Heat and energy are fundamental concepts that drive the motion and transformations of matter in our universe. From the warmth of a summer day to the power of engines propelling vehicles, the interplay between heat and energy is at the heart of countless natural processes and human technologies. Exploring these concepts unravels the intricate mechanisms behind motion, change, and the dynamic forces that shape our world.

Heat and Temperature: A Matter of Motion

Heat is the transfer of thermal energy between objects due to a temperature difference. Temperature is a measure of the average kinetic energy—the movement—of particles within a substance. Hotter objects have higher kinetic energy, and heat flows from hotter to cooler objects.

Thermodynamics: The Laws Governing Energy

Thermodynamics is the study of heat and energy transformations. The laws of thermodynamics establish fundamental principles:

1. **First Law (Law of Conservation of Energy)**: Energy cannot be created or destroyed, only converted from one form to another.

2. **Second Law**: Heat naturally flows from hot to cold objects. It's difficult to convert all heat into work, as some is always lost to less useful forms.

3. **Third Law**: As temperature approaches absolute zero (-273.15°C or 0 Kelvin), entropy (a measure of disorder) approaches a minimum value.

Forms of Energy: A Spectrum of Transformations

Energy comes in various forms:

1. **Mechanical Energy**: The energy of motion (kinetic) and stored energy (potential) due to an object's position.
2. **Thermal Energy**: The total kinetic energy of particles within a substance.
3. **Chemical Energy**: Stored energy in chemical bonds, released during reactions.
4. **Electromagnetic Energy**: Energy carried by electromagnetic waves, including visible light, radio waves, and more.

Heat Transfer: The Movement of Thermal Energy

Heat transfer occurs through three mechanisms:

1. **Conduction**: Heat transfer through direct contact of particles, such as a metal spoon getting hot when placed in hot soup.
2. **Convection**: Heat transfer through the movement of fluids (liquids or gases), like warm air rising and cold air sinking.
3. **Radiation**: Heat transfer through electromagnetic waves, like the Sun's energy reaching Earth.

Thermal Expansion: Objects Respond to Heat

As substances heat up, their particles move faster, leading to expansion. This phenomenon has practical applications, from the design of bridges to the functioning of thermostats.

Heat Engines and Work: Harnessing Energy

Heat engines, like car engines and power plants, convert thermal energy into mechanical work. The efficiency of a heat engine depends on the temperature difference between the hot and cold reservoirs.

Entropy and the Arrow of Time: The Quest for Order

Entropy, a measure of the dispersal of energy in a system, tends to increase over time. This principle gives rise to the arrow of time, the irreversibility of natural processes.

Renewable and Non-renewable Energy: Powering Our World

Energy sources are classified as renewable (solar, wind, hydroelectric) or non-renewable (fossil fuels). As concerns about environmental impact and resource depletion grow, the focus shifts toward sustainable energy options.

Energy Efficiency: Maximizing Resources

Energy efficiency involves reducing energy waste and optimizing energy use. It's crucial for conserving resources, reducing emissions, and minimizing costs.

Conclusion: Energizing Our Understanding

The concepts of heat and energy are the driving forces behind the dynamic world we inhabit. From the cycles of nature to the marvels of modern technology, heat and energy dictate how matter moves, transforms, and sustains life. Understanding these principles empowers us to harness the potential of energy sources responsibly, shaping a future where sustainable energy solutions play a pivotal role in the well-being of our planet and its inhabitants.

CHAPTER 31: FORCES TOGETHER: HUNTING FOR A SUPER RULE

The quest for understanding the fundamental nature of our universe has led scientists on a journey to uncover the laws that govern its behavior. Among these pursuits is the search for a unifying theory—a "super rule"—that can explain and harmonize the diverse forces that shape the cosmos. From gravity's pull to the electromagnetic interactions that power our devices, the endeavor to unite these forces reveals the beauty and complexity of the universe's underlying structure.

The Forces at Play: A Cosmic Choreography

Four fundamental forces are recognized in the universe:

1. **Gravity**: The force that governs the motion of celestial bodies, from planets orbiting stars to galaxies in cosmic dance.

2. **Electromagnetic Force**: The force responsible for interactions between electrically charged particles, enabling everything from chemical reactions to the functioning of electronic devices.

3. **Weak Nuclear Force**: This force governs certain types of radioactive decay, influencing the behavior of subatomic particles.

4. **Strong Nuclear Force**: This powerful force binds protons and neutrons together in atomic nuclei, giving rise to the stability of matter.

The Quest for Unification: The Grand Unified Theory

Scientists have made significant strides in understanding the fundamental forces by seeking ways to unify them. One step is the pursuit of a Grand Unified Theory (GUT), which aims to merge the strong nuclear force with the weak and electromagnetic forces.

Beyond GUT: The Theory of Everything

An even loftier ambition is the Theory of Everything (TOE), a hypothetical framework that would encompass all fundamental forces and particles in a single, coherent description. String theory and its variants are among the leading candidates for a TOE, proposing that particles are not point-like, but rather tiny, vibrating strings.

String Theory: A Symphony of Vibrations

String theory suggests that particles are not truly "particles" but rather tiny, oscillating strings of energy. These vibrations give rise to the diverse particles we observe, and the different patterns of vibrations correspond to different particle properties.

Supersymmetry: A Link Between Particles and Forces

Supersymmetry is a theoretical framework that proposes a symmetry between particles and forces, suggesting that each known particle has a yet-to-be-discovered superpartner particle. This symmetry could provide a deeper understanding of the forces' unification.

Quantum Gravity: Bridging the Macro and Micro

A significant challenge in unifying forces is reconciling the laws of gravity with the principles of quantum mechanics. Quantum gravity theories, like loop quantum gravity and string theory, aim to merge these two fundamental frameworks.

Cosmic Signatures: The LHC and Beyond

Particle accelerators like the Large Hadron Collider (LHC) at CERN allow scientists to probe the fundamental forces at energy scales that mimic the early universe. The discovery of the Higgs boson at the LHC was a significant step in understanding the origin of mass and the electroweak force unification.

The Mysteries Persist: The Beauty of Exploration

While progress has been made in understanding the fundamental forces, many mysteries remain. Dark matter, dark energy, and the exact nature of the gravitational force are among the enigmas that continue to drive scientific inquiry.

Conclusion: The Cosmic Symphony

The journey to uncover a super rule that unites the forces of the universe is a testament to human curiosity and ingenuity. It reflects our innate desire to unravel the intricacies of the cosmos and glimpse the underlying harmony that governs its behavior. While the super rule remains elusive, the pursuit of unification transcends scientific boundaries, offering a glimpse into the interconnected nature of the universe and the endless possibilities that lie beyond our current understanding.

CHAPTER 32: TALKING BITS: HOW WE SHARE INFORMATION

The exchange of information is a cornerstone of human civilization, enabling us to convey ideas, emotions, and knowledge across time and space. In an increasingly interconnected world, the methods we use to communicate have evolved dramatically, from spoken language to digital signals. The concept of "talking bits" encompasses the diverse ways in which we share information, bridging gaps and fostering connections in our global society.

Language and Communication: A Uniquely Human Trait

Language, both spoken and written, is the most intricate and versatile form of communication. It allows us to convey abstract concepts, emotions, and complex ideas. The evolution of language has been central to the development of culture, technology, and societal norms.

Non-Verbal Communication: Beyond Words

Non-verbal cues, including body language, facial expressions, and gestures, play a vital role in communication. They provide context, emotional nuances, and supplementary information to spoken or written words.

Visual Communication: The Power of Images

Visual communication employs images, graphics, and visual elements to convey information. It is a universal language that transcends cultural and linguistic barriers, making it effective for disseminating information to diverse audiences.

Digital Communication: The Information Age

The digital revolution has transformed how we share information:

1. **Internet**: The World Wide Web connects people globally, facilitating instant access to vast amounts of information.

2. **Social Media**: Platforms like Facebook, Twitter, and Instagram have revolutionized how we share updates, opinions, and images with friends, family, and the world.

3. **Email**: Electronic mail has transformed written correspondence, making communication faster and more convenient.

4. **Instant Messaging**: Apps like WhatsApp and Messenger enable real-time text, voice, and video communication across the globe.

Mass Media: Broadcasting to the Masses

Mass media, including television, radio, and newspapers, have long been essential for disseminating information to wide audiences. They play a significant role in shaping public opinion and cultural trends.

Multimodal Communication: The Fusion of Forms

Modern communication often combines various modes:

1. **Multimedia Presentations**: Slideshows and videos enhance spoken or written content, catering to different learning styles.

2. **Podcasts**: Audio shows cover a wide range of topics, providing a platform for in-depth discussions and storytelling.

3. **Infographics**: Visual representations of data and information make complex concepts more accessible.

Globalization and Language Diversity: Bridging Cultures

In our interconnected world, people with different native languages need tools to communicate effectively. Translation apps, language courses, and global communication platforms bridge linguistic barriers.

Challenges and Considerations: Navigating the Digital Age

Digital communication comes with challenges:

1. **Misinformation**: The rapid spread of unverified or false information can lead to misunderstandings and public health risks.

2. **Privacy Concerns**: Digital platforms raise questions about the security and privacy of personal information.

3. **Digital Divide**: Unequal access to technology can create disparities in information sharing and connectivity.

The Future of Communication: Evolving Landscape

As technology continues to advance, the future of communication holds exciting possibilities:

1. **Artificial Intelligence**: AI-driven language translation and communication tools promise to further break down language barriers.

2. **Virtual Reality**: Immersive experiences could revolutionize how we interact and communicate with each other.

3. **Quantum Communication**: Quantum technology offers secure and ultra-fast communication methods.

Conclusion: The Symphony of Sharing

"Talking bits" encapsulates the diverse and dynamic ways we share information. From the earliest human conversations to the vast digital networks of today, communication has shaped societies, driven innovation, and facilitated connections across borders and cultures. As we navigate this intricate landscape, understanding the nuances of language, media, and technology empowers us to communicate effectively, build bridges, and harness the power of information to shape a more informed and connected global community.

CHAPTER 33: CIRCLE OF LIFE: BIRTH, GROWTH, AND SAYING GOODBYE

The cycle of life, an eternal rhythm that threads through existence, encompasses the profound journey from birth to growth, and eventually, the poignant act of saying goodbye. It's a tapestry woven with the threads of beginnings, transformations, and endings, illuminating the intricate dance of existence and the profound lessons it imparts. From the emergence of new life to the inevitable departure, the circle of life encapsulates the beauty, challenges, and universal truths that define our human experience.

Birth: The Dawn of Life

Birth heralds the commencement of life's journey—a fragile, wondrous moment when a new being enters the world. This miracle is a testament to the resilience and adaptability of life, evoking awe and celebration as a tiny spark joins the vast canvas of existence.

Growth and Development: Navigating the Passage of Time

From infancy to adulthood, growth and development shape our experiences. The cycle of life sees the emergence of personalities, talents, and dreams. Learning, maturing, and adapting to the world are essential stages that allow individuals to carve their unique paths.

Human Connections: The Tapestry of Relationships

As individuals grow, they weave intricate relationships with others, creating a vibrant tapestry of human connections. Friends, family, mentors, and companions shape our journey, offering support, love, and shared experiences that enrich our lives.

Cultural Perspectives: Varying Traditions and Beliefs

Cultures around the world approach the circle of life with diverse rituals and beliefs. Birth and coming-of-age ceremonies, weddings, and rites of passage honor these transitions, underscoring the shared human experience that transcends borders.

Saying Goodbye: Navigating the Cycle's End

Saying goodbye, the final act of the circle, carries profound emotional weight. Whether through the natural course of life or unexpected circumstances, facing mortality prompts contemplation and reflection on the essence of existence.

Grief and Loss: The Process of Healing

Grief is a complex emotional response to loss, encompassing stages of denial, anger, bargaining, depression, and acceptance. Navigating grief is a deeply personal journey, marked by introspection, support from loved ones, and the gradual mending of a wounded heart.

Legacy and Remembrance: Immortality Through Memory

In the wake of farewells, legacies endure—testaments to a life lived. Through achievements, stories, and the indelible impact on others, individuals leave imprints that transcend their earthly presence.

Spiritual and Philosophical Reflections: Seeking Meaning

Throughout history, philosophers, thinkers, and spiritual leaders have pondered the nature of life's cycle. Concepts of reincarnation, resurrection, and the continuity of the soul offer perspectives on the continuum of existence.

Nature's Cycles: Echoes of Life's Journey

Observing the cycles of nature—seasons, birth, growth, decay, and rebirth—mirrors the human experience. Nature's rhythms offer solace and wisdom, reminding us that the circle of life is a universal pattern woven into the fabric of reality.

The Transcendent Narrative: The Universal Human Experience

The circle of life transcends cultures, beliefs, and time periods, binding humanity in a universal narrative. It is a reminder of our shared vulnerability, aspirations, and the undeniable connection that unites us all.

Conclusion: Embracing the Wholeness of Existence

The circle of life is an exquisite dance of beginnings and endings, growth and transformation, joy and sorrow. It's a reminder that life is both fleeting and precious, and every moment is an opportunity to contribute to the grand symphony of existence. Through birth, growth, and saying goodbye, we embrace the beauty and complexity of the human journey, understanding that every phase is an essential note in the melody that defines our existence on this intricate and awe-inspiring planet.

CHAPTER 34: MIXING THINGS UP: CHEMISTRY AT WORK

Chemistry is the intricate language of matter—the science that explains how substances interact, transform, and combine to create the world around us. From the colors of a sunrise to the composition of a complex molecule, chemistry is the fundamental force that underpins the physical and chemical changes that shape our universe. Delving into the topic of "Mixing Things Up" unveils the mesmerizing dance of atoms and molecules, offering insights into the reactions, compounds, and materials that define our everyday lives.

The Atom's Dance: Building Blocks of Matter

Atoms are the building blocks of matter, each with a unique set of properties defined by its atomic number, mass, and arrangement of subatomic particles—protons, neutrons, and electrons.

Chemical Reactions: A Symphony of Transformation

Chemical reactions are the essence of chemistry, where atoms rearrange to form new substances. These reactions follow the principles of conservation of mass and energy, revealing the delicate balance of matter's transformations.

Chemical Bonds: The Glue of Matter

Chemical bonds—ionic, covalent, and metallic—bind atoms together into molecules and compounds. These bonds dictate the properties of substances, from the strength of metals to the electrical conductivity of saltwater.

States of Matter: Matter in Motion

Matter exists in different states—solid, liquid, gas, and plasma—depending on the arrangement and motion of its particles. Phase transitions, like melting, boiling, and sublimation, are governed by energy changes.

Mixing and Solutions: The Dance of Particles

Mixtures and solutions are the result of combining different substances. Solutions involve solutes dissolving in solvents, leading to a homogenous mixture. Understanding concentration, saturation, and solubility is crucial in chemistry.

Chemical Elements and the Periodic Table: Organizing Matter

The periodic table arranges elements based on their properties and atomic numbers. It reveals patterns in element behavior and provides insights into the relationships between different elements.

Acids and Bases: The pH Scale

Acids and bases are fundamental concepts in chemistry, influencing reactions, solutions, and everyday products. The pH scale measures the acidity or basicity of substances.

Organic Chemistry: The Chemistry of Carbon

Organic chemistry focuses on carbon-containing compounds, the basis of life. Organic molecules are diverse and complex, with vital roles in biology, industry, and medicine.

Materials Science: Engineering New Possibilities

Materials science explores the properties, structure, and applications of materials. Advances in materials science lead to innovations in technology, medicine, and manufacturing.

Chemistry in the Environment: The Global Impact

Chemistry plays a role in environmental challenges, from air and water pollution to climate change. Understanding chemical processes is essential for developing sustainable solutions.

Biochemistry: Chemistry of Life

Biochemistry studies the chemical processes within living organisms. It explores topics like enzymes, metabolism, and the molecular basis of genetic information.

Nanotechnology: The World of the Extremely Small

Nanotechnology manipulates matter at the nanoscale, leading to innovations in medicine, electronics, and materials science.

Chemistry and Everyday Life: The Invisible Threads

Chemistry is woven into our daily lives, from cooking and cleaning to personal care products and technology. It shapes the materials we use and the energy we consume.

Conclusion: Unveiling the Molecular Ballet

"Mixing Things Up" in the realm of chemistry reveals the unseen choreography of molecules—their interactions, transformations, and endless possibilities. It's a journey that empowers us to understand, harness, and shape the world around us. From the smallest particle to the grandest cosmic phenomenon, chemistry's influence is boundless, inviting us to explore the mysteries of matter, the elegance of reactions, and the extraordinary impact of understanding how things mix, transform, and interact in the magnificent tapestry of our universe.

CHAPTER 35: THINKING MACHINES: HOW OUR BRAIN WORKS

The human brain, an intricate web of neurons, synapses, and electrical signals, is one of the most fascinating and complex structures in the known universe. It holds the key to our thoughts, emotions, memories, and consciousness. Understanding how our brain works is an ongoing journey that spans the realms of neuroscience, psychology, biology, and philosophy, offering insights into the nature of intelligence, creativity, and the essence of being human.

Neurons: The Messengers of the Mind

Neurons are the fundamental units of the brain, specialized cells that transmit information through electrical and chemical signals. These cells connect to each other through synapses, forming intricate neural networks that enable communication and processing.

Brain Structure: Orchestrating Complexity

The brain is composed of different regions, each responsible for specific functions:

1. **Cerebral Cortex**: This outer layer is crucial for conscious thought, reasoning, perception, and decision-making.

2. **Limbic System**: Linked to emotions and memory, it plays a key role in shaping our reactions and experiences.

3. **Brainstem and Cerebellum**: These regions control essential functions like breathing, heart rate, and balance.

Neurotransmitters: The Chemical Messengers

Neurotransmitters are chemicals that transmit signals between neurons across synapses. They influence mood, behavior, cognition, and more. Imbalances can lead to mental health disorders.

Electrical Signaling: The Language of Neurons

Neurons communicate through electrical signals known as action potentials. These rapid changes in voltage enable information to travel across neural pathways.

Plasticity: The Brain's Adaptive Nature

Neuroplasticity is the brain's ability to reorganize and adapt. It underpins learning, memory, and recovery from brain injuries.

Memory and Learning: Unveiling the Enigma

Memory is a multifaceted process involving encoding, storage, and retrieval. Long-term potentiation—the strengthening of synapses—plays a vital role in learning and memory formation.

Consciousness and Perception: The Mind's Theatre

Consciousness encompasses our awareness of self and surroundings. Perception, the brain's interpretation of sensory information, shapes our reality and subjective experiences.

Emotions and Decision-Making: The Heart of Humanity

The brain's limbic system plays a central role in generating emotions. Decision-making involves complex interplay between rational and emotional processes.

Creativity and Imagination: Unleashing the Mind's Potential

Creativity arises from the brain's ability to connect seemingly unrelated ideas, forming novel solutions and innovations.

Disorders and Disorders: Unraveling the Mind's Mysteries

Mental health disorders, from depression and anxiety to schizophrenia, have complex origins involving genetics, brain chemistry, and environmental factors. Research aims to understand and treat these conditions.

Technological Advances: Mapping the Mind

Brain imaging techniques like fMRI and EEG allow scientists to observe brain activity and structure. The Human Connectome Project aims to map neural connections, shedding light on brain networks.

Artificial Intelligence and Brain-Computer Interfaces: Bridging the Gap

Advances in AI and brain-computer interfaces hold the promise of enhancing human cognition, communication, and mobility.

The Hard Problem of Consciousness: A Philosophical Quandary

The nature of consciousness—how subjective experiences emerge from physical processes—is a philosophical puzzle that intertwines neuroscience, psychology, and philosophy.

Ethical Considerations: Exploring Boundaries

Understanding the brain raises ethical questions about cognitive enhancement, privacy, and the potential manipulation of thoughts and emotions.

Conclusion: The Universe Within

The study of how our brain works is a voyage into the depths of human existence. It unravels the marvels of cognition, emotion, and consciousness—the very essence of what it means to be alive and self-aware. As we explore the intricate dance of neurons and synapses, we gain insights into our unique abilities, the mechanisms that shape our perceptions, and the awe-inspiring complexity of the human experience that resides within the folds and convolutions of our extraordinary brains.

CHAPTER 36: SIMPLE MACHINES: TOOLS WE USE EVERY DAY

Simple machines are the unsung heroes of our daily lives—underappreciated yet indispensable tools that make work easier by multiplying force or changing the direction of a force. From the wheels that propel vehicles to the pulleys that lift heavy loads, simple machines are the building blocks of more complex devices. Understanding these fundamental tools unveils the elegant principles of physics that underlie our technological advancements and everyday conveniences.

The Six Simple Machines: A Foundation of Mechanics

There are six basic types of simple machines:

1. **Lever**: A rigid bar that pivots around a fixed point, or fulcrum. Examples include seesaws, scissors, and crowbars.

2. **Wheel and Axle**: A circular object (wheel) attached to a rod (axle) that enables rotational motion. Examples include wheels, doorknobs, and rolling pins.

3. **Pulley**: A wheel with a groove along its edge that a rope or chain passes through. Pulleys change the direction of a force and can create mechanical advantage. Elevators and flagpoles often use pulleys.

4. **Inclined Plane**: A flat surface that is slanted, allowing for easier movement of objects. Ramps, chisels, and screws are examples of inclined planes.

5. **Wedge**: A triangular-shaped tool that separates objects by applying force in opposite directions. Wedges are found in tools like knives, axes, and nails.

6. **Screw**: An inclined plane wrapped around a cylinder.

Screws are used to hold objects together or create linear motion, as seen in bolts, jar lids, and drills.

Mechanical Advantage: Making Work Easier

Simple machines provide mechanical advantage—the ability to multiply force or change the direction of a force. By using these machines, we can achieve tasks that would be difficult or impossible without them.

Combining Machines: The Power of Synergy

Complex machines are often combinations of simple machines. For instance, bicycles use wheels and axles, levers, and inclined planes. Cars employ numerous simple machines, such as gears and pulleys, in their complex systems.

Practical Applications: From Ancient Times to Modern Technology

Throughout history, simple machines have been pivotal in technological advancements:

1. **Ancient Egypt**: Pulleys aided in constructing the pyramids.

2. **Industrial Revolution**: Machines powered by simple machines led to significant advancements in manufacturing and transportation.

3. **Modern Era**: Technology relies on simple machines —gears in machinery, wheels in transportation, and pulleys in cranes.

Teaching Physics Concepts: A Gateway to Learning

Understanding simple machines provides an accessible entry point to physics concepts such as force, work, energy, and mechanical advantage. It lays the foundation for more complex studies in engineering and applied physics.

Engineering Innovations: Building on Simplicity

Engineers harness the principles of simple machines to

create intricate designs. Robotics, medical devices, and space exploration owe their advancements to the fundamental understanding of these tools.

Education and Exploration: Fostering Curiosity

Studying simple machines encourages curiosity, critical thinking, and hands-on exploration. Educational kits, DIY projects, and interactive experiences allow learners of all ages to grasp physics principles in a tangible way.

Conclusion: The Unsung Heroes

Simple machines are the unassuming architects of our modern world. From the monumental achievements of ancient civilizations to the intricacies of cutting-edge technology, simple machines remain the bedrock upon which innovation is built. Their simplicity belies their impact, as they quietly enhance our lives, empower engineers, and serve as the foundational pieces that enable us to overcome challenges, create marvels, and keep the wheels of progress turning.

CHAPTER 37: FAMILY TRAITS: HOW WE INHERIT TRAITS

The tapestry of human diversity is woven from a complex interplay of genetic information passed down through generations. Family traits, the unique characteristics that define us, are a result of a fascinating journey involving inheritance, genetics, and the interaction between our genes and the environment. Delving into the realm of family traits unveils the intricate mechanisms that shape our physical features, behaviors, and susceptibilities, offering a glimpse into the incredible complexity of life's blueprint.

Genes and Heredity: Unveiling the Blueprint

Genes are segments of DNA that contain the instructions for building and maintaining an organism. These genetic instructions guide the development of traits—physical, physiological, and even behavioral.

Dominant and Recessive Traits: The Rule of Inheritance

Each individual inherits two copies of each gene, one from each parent. Dominant traits mask the effects of recessive traits. A dominant gene needs only one copy to manifest, while two copies of a recessive gene are required.

Punnett Squares: Predicting Inheritance

Punnett squares are tools that predict the probability of offspring inheriting specific traits from their parents. These visual aids help us understand how different combinations of genes lead to a range of outcomes.

Mendelian Genetics: Peas and Principles

Gregor Mendel's work with pea plants in the 19th century laid the foundation for modern genetics. His observations of traits' inheritance patterns established the laws of segregation and independent assortment.

Complex Traits: Beyond Simple Inheritance

Some traits, like height or intelligence, are influenced by multiple genes and environmental factors. These complex traits do not follow simple dominant-recessive patterns and exhibit greater variability.

Genetic Variation: The Spice of Life

Genetic diversity arises from mutations, the random changes in DNA sequences. While some mutations lead to genetic disorders, others contribute to evolution, adaptation, and the spectrum of traits we observe.

Chromosomes and Sex-Linked Traits: A Gendered Influence

Sex-linked traits are associated with genes located on the X or Y chromosomes. These traits often exhibit different patterns of inheritance in males and females due to the presence of only one X chromosome in males.

Epigenetics: Beyond Genetics

Epigenetics explores how environmental factors can influence gene expression. Chemical modifications to DNA and associated proteins can be inherited and impact traits even without changes to the underlying DNA sequence.

Inheritance in Families: The Generational Threads

Family traits are a result of genetic inheritance across generations. Observing shared traits among family members can provide insights into genetic legacies.

Nature vs. Nurture: The Dueling Forces

The debate between nature (genetics) and nurture (environment) continues to shape our understanding of how traits develop. While genes set the foundation, environmental factors play a vital role in trait expression.

Genetic Counseling and Testing: Empowering Choices

Genetic counseling helps individuals understand their risk of inheriting or passing on genetic conditions. Genetic testing provides insights into potential health concerns, family history, and ancestry.

Ethical Considerations: Navigating Genetic Knowledge

As genetic knowledge expands, ethical dilemmas arise concerning privacy, informed consent, and potential discrimination based on genetic information.

Future Possibilities: Unraveling the Genetic Code

Advances in genetics, like gene editing technologies (CRISPR), offer potential to correct genetic disorders and shape future generations.

Conclusion: The Intricacies of Inheritance

Family traits embody the legacy of generations, a testament to the intricate dance of genes and environment. From the simplest inherited traits to the complex interplay of genes and factors that define our identities, family traits are a window into the remarkable complexity of life itself. As we unravel the secrets of inheritance, we gain not only a deeper understanding of our individual uniqueness but also an appreciation for the grand tapestry of human diversity that emerges from the interweaving threads of genetic information passed down through time.

CHAPTER 38: LIGHT AND MORE: EXPLORING DIFFERENT RADIATIONS

Radiation is a fundamental phenomenon that traverses the realms of physics, chemistry, biology, and cosmology. From the vibrant colors of the visible spectrum to the invisible waves that power our technology, exploring different radiations unveils the multifaceted nature of energy and its impact on the universe. This journey delves into the spectrum of radiations, from the familiar glow of light to the exotic cosmic rays that traverse the cosmos, illuminating the profound connections between matter, energy, and the intricacies of our reality.

Electromagnetic Radiation: The Spectrum Unveiled

Electromagnetic radiation is energy propagated through oscillating electric and magnetic fields. The electromagnetic spectrum encompasses a vast range of wavelengths and frequencies, each with distinct properties and applications.

Visible Light: The Colors of Perception

Visible light, the part of the spectrum perceivable by the human eye, manifests as a stunning array of colors. Its interactions with matter give rise to reflection, refraction, and diffraction, creating the world of colors we experience.

Infrared Radiation: Warmth and Beyond

Infrared radiation is heating energy emitted by objects due to their temperature. Its applications span from thermal imaging and remote sensing to communication and even cooking.

Ultraviolet Radiation: The Enigma of Sunlight

Ultraviolet (UV) radiation lies beyond violet in the visible spectrum. While UV rays from the sun can be harmful, they also play a vital role in processes like vitamin D synthesis and sterilization.

X-rays: Peering Inside Matter

X-rays are high-energy electromagnetic waves capable of penetrating matter. They are used in medical imaging, materials testing, and studying the structure of crystals and molecules.

Gamma Rays: Cosmic Powerhouses

Gamma rays are the most energetic form of electromagnetic radiation, often originating from nuclear reactions and cosmic phenomena. They play a critical role in understanding the cosmos and probing the fundamental nature of matter.

Radio Waves: Messages from Afar

Radio waves have the longest wavelengths in the electromagnetic spectrum. They power communication systems, including radio broadcasts, television, cell phones, and wireless networks.

Microwaves: From Popcorn to the Universe

Microwaves have wavelengths longer than infrared but shorter than radio waves. They are used in microwave ovens, radar systems, and cosmology, where they provide insights into the early universe.

Cosmic Rays: High-Energy Particles from Space

Cosmic rays are high-energy particles from outer space that impact the Earth's atmosphere. They hold clues about distant astrophysical processes and contribute to the ionization of the atmosphere.

Quantum Mechanics and Radiations: A Subatomic Dance

Quantum mechanics describes how radiations interact with matter on the subatomic scale. The photoelectric effect, for example, led to the understanding of light as discrete packets of energy called photons.

Applications and Innovations: Harnessing Radiations

Radiations have led to numerous technological advancements:

1. **Medical Imaging**: X-rays, CT scans, and PET scans offer insights into the human body's internal structures.

2. **Communication**: Radio waves, microwaves, and light form the backbone of modern communication systems.

3. **Energy Generation**: Solar panels convert sunlight into electricity.

4. **Materials Analysis**: Spectroscopy uses radiations to analyze the composition of materials.

5. **Astrophysics**: Telescopes observe radiations from cosmic sources, unveiling the universe's mysteries.

Safety and Ethical Considerations: The Balancing Act

While radiations have brought tremendous benefits, their potential hazards require careful management and regulations. Radiation safety, medical ethics, and environmental impact are crucial considerations.

Conclusion: A Kaleidoscope of Energy

Exploring different radiations is a voyage through the spectrum of energy that permeates our world and the cosmos. From the microscopic dance of subatomic particles to the grand cosmic ballet, radiations shape our understanding of matter, light, and the fundamental nature of reality. Whether revealing the secrets of the human body or unraveling the mysteries of distant galaxies, radiations empower us to grasp the intricate connections between energy, matter, and the universe at large, offering a glimpse into the harmonious interplay of forces that shapes our existence.

CHAPTER 39: INSIDE YOU: HOW YOUR BODY WORKS

The human body is a remarkable symphony of interconnected systems, organs, and cells working in harmony to sustain life. From the intricate dance of enzymes in our cells to the rhythmic beats of the heart, understanding how your body works unveils the complex biological processes that enable us to breathe, move, think, and thrive. This journey explores the wonders of human anatomy and physiology, delving into the intricate mechanisms that keep us alive, healthy, and in tune with the world around us.

Cells: The Building Blocks of Life

Cells are the smallest units of life, each with specialized functions. They form tissues, organs, and systems, working together to maintain the body's balance.

Tissues and Organs: A Symphony of Function

Tissues—groups of similar cells—combine to create organs. These organs, such as the heart, lungs, and brain, collaborate to perform specific functions critical to survival.

Skeletal System: The Body's Foundation

The skeletal system provides structure, support, and protection. Bones are dynamic structures that grow, reshape, and regenerate over time.

Muscular System: Powering Movement

Muscles allow movement, from the precise contractions of facial muscles to the powerful actions of the legs. Skeletal, smooth, and cardiac muscles contribute to various bodily functions.

Cardiovascular System: The Rhythmic Pump

The heart and blood vessels make up the cardiovascular

system. The heart pumps blood, carrying oxygen and nutrients throughout the body while removing waste.

Respiratory System: Breathing Life

The respiratory system facilitates the exchange of oxygen and carbon dioxide. The lungs expand and contract, enabling us to inhale life-sustaining oxygen and exhale waste gases.

Digestive System: Nourishing the Body

The digestive system breaks down food, extracting nutrients for energy, growth, and repair. From the mouth to the intestines, this system plays a vital role in maintaining health.

Nervous System: Mastering Communication

The nervous system coordinates bodily functions and responses. The brain, spinal cord, and nerves transmit electrical signals that control movements, thoughts, and emotions.

Endocrine System: Hormones and Harmony

The endocrine system uses hormones to regulate various bodily processes, from growth and metabolism to mood and reproduction.

Immune System: Guardians of Health

The immune system defends the body against harmful invaders like bacteria, viruses, and fungi. It includes a network of cells, tissues, and organs working together to maintain health.

Integumentary System: The Body's Shield

The skin, hair, and nails form the integumentary system. Skin acts as a barrier against pathogens, regulates temperature, and houses sensory receptors.

Reproductive System: The Cycle of Life

The reproductive system enables the perpetuation of the species. In females, it includes the ovaries, uterus, and mammary glands, while males have testes and other structures.

Homeostasis: Balancing Act of Life

Homeostasis is the body's ability to maintain a stable internal environment despite external changes. It ensures optimal conditions for bodily functions.

Health and Disease: Striking a Balance

Health is a state of physical, mental, and social well-being. Diseases result from disruptions in normal bodily functions and can be caused by pathogens, genetic factors, lifestyle choices, and environmental influences.

Medical Advances: Unveiling Mysteries

Scientific breakthroughs have illuminated the intricacies of the human body, leading to innovative treatments, medications, and medical technologies.

Lifestyle and Wellness: Nurturing Your Body

Maintaining a balanced diet, regular exercise, adequate sleep, and managing stress are vital for overall wellness.

Ethical Considerations: The Human Body's Boundaries

Medical ethics, informed consent, and debates surrounding topics like organ transplantation and genetic engineering are critical ethical considerations tied to our understanding of the human body.

Conclusion: A Living Symphony

Understanding how your body works is a journey into the marvels of life itself. From the microcosm of cells to the intricate orchestration of organ systems, the human body is a testament to the intricate dance of biology, chemistry, and physics that define our existence. As we delve into the beauty and complexity of our inner workings, we gain not only a deeper appreciation for the miracle of life but also an awareness of the delicate balance that sustains us—a symphony of rhythms and processes that harmonize to create the incredible journey of being alive.

CHAPTER 40: QUANTUM WONDERS: STRANGE SCIENCE OF THE TINY

Quantum physics, often called quantum mechanics, is a realm of science that delves into the bizarre and mind-bending behaviors of the tiniest particles that make up our universe. From the curious dance of particles at the subatomic level to the perplexing phenomena that challenge our understanding of reality, exploring the quantum wonders unveils a world of strangeness that defies classical intuition. This journey takes us through the fundamental principles, experiments, and mysteries that make up the strange science of the quantum realm.

The Quantum Revolution: A Paradigm Shift

Quantum physics emerged in the early 20th century as a response to the limitations of classical physics. It revolutionized our understanding of matter, energy, and the very fabric of reality.

Wave-Particle Duality: A Paradox Unveiled

Quantum theory introduced the concept of wave-particle duality. Particles like electrons and photons can behave as both waves and particles, blurring the line between classical categories.

Quantum Superposition: A State of Both-And

Superposition is a fundamental quantum phenomenon where particles can exist in multiple states simultaneously. This bizarre principle underpins quantum computing and quantum cryptography.

Entanglement: Spooky Action at a Distance

Entanglement is perhaps one of the most mind-bending aspects of quantum physics. When particles become entangled, their states become correlated, no matter the distance between them.

Changes in one particle's state instantaneously affect the other, defying classical notions of causality.

Quantum Uncertainty: The Heisenberg Principle

The Heisenberg Uncertainty Principle states that there's a fundamental limit to how precisely we can know both the position and momentum of a particle. This introduces inherent uncertainty into the very fabric of the quantum world.

Quantum Mechanics and Reality: The Observer Effect

The act of observing a quantum system can change its behavior. This has led to philosophical debates about the nature of reality and the role of consciousness in shaping the quantum world.

Quantum Tunneling: Breaking Barriers

Quantum tunneling allows particles to pass through energy barriers that, according to classical physics, should be impenetrable. It's a phenomenon crucial to understanding nuclear fusion, semiconductors, and more.

Quantum Computing: Beyond Classical Limits

Quantum computers leverage superposition and entanglement to perform certain calculations exponentially faster than classical computers. This promises breakthroughs in cryptography, optimization, and simulation.

Quantum Reality and Interpretations: Many Worlds, Many Debates

Interpretations of quantum mechanics abound, each offering a unique perspective on how to understand the strange phenomena. The Copenhagen interpretation, Many-Worlds theory, and pilot-wave theory are just a few examples.

Applications and Technological Implications: Quantum Leap

Quantum technologies are emerging in fields like cryptography, secure communication, precision measurements, and materials science. Quantum computers hold the potential to revolutionize various industries.

Challenges and Open Questions: The Quest Continues

Despite the remarkable progress, quantum physics grapples with unanswered questions—such as reconciling quantum mechanics with general relativity and the elusive nature of dark matter and dark energy.

Philosophical Implications: The Nature of Reality

Quantum physics has profound philosophical implications, challenging our understanding of causality, determinism, and the very nature of existence.

Ethical Considerations: Quantum Ethics

As quantum technologies advance, ethical considerations emerge, such as the impact on cybersecurity, privacy, and the potential for quantum weapons.

Conclusion: The Enigma of the Quantum World

Quantum wonders take us on a journey into the fabric of reality itself—a realm where particles defy classical boundaries and exhibit behaviors that stretch the limits of imagination. This strange science has upended our understanding of the universe and has the potential to reshape the way we think, compute, and communicate. As we continue to probe the depths of the quantum realm, we unravel the enigma of the tiniest particles that weave the tapestry of existence, inviting us to embrace the awe-inspiring beauty and mystery of the quantum world.

EPILOGUE

As we conclude our journey through the pages of "The Everyday Science: Where Curiosity and Reality Converge," we find ourselves at the crossroads of wonder and understanding. The tapestry of scientific exploration that we've woven together has illuminated the intricacies of our world, revealing both its beauty and its complexity. From the tiniest particles that make up matter to the grand expanses of the cosmos, we've explored the spectrum of existence, delving into the questions that have shaped human understanding for generations.

This book has been a celebration of curiosity—an invitation to peer beneath the surface of the ordinary and uncover the extraordinary. We've witnessed the magic of water, the dance of elements, and the profound impact of microbes. We've marveled at the interplay of colors and light, grappled with the forces that guide our universe, and ventured into the realm of quantum wonders.

But our exploration doesn't end here. Science is a never-ending journey—an ongoing quest to explore the unknown, to challenge assumptions, and to peel back the layers of reality to reveal deeper truths. The mysteries we've encountered are not barriers, but invitations to keep seeking, to keep questioning, and to keep pushing the boundaries of what we know.

Each discovery, each insight, and each revelation are a stepping stone on this path of exploration. Our understanding deepens, our perspectives broaden, and our sense of wonder continues to expand. As we look back at the myriad wonders we've encountered, we realize that the universe is a canvas of infinite possibilities—a canvas waiting for us to add our own strokes of curiosity, inquiry, and discovery.

As we close this chapter, let us carry forward the spirit of inquiry that has guided us through the intricacies of everyday science. Let us look at the world with new eyes, seeing not just the surface but the stories that lie beneath. Let us embrace the beauty of complexity and the elegance of simplicity, recognizing that in every corner of existence, there is a wealth of knowledge waiting to be uncovered.

The journey of science is a journey of the mind, the heart, and the human spirit. It's a journey that connects us to the past, enriches our present, and shapes the future. It's a journey that has transformed the way we perceive ourselves, our world, and the universe that envelops us.

So, as we close this book, let us remember that the quest for understanding is never truly complete. It's a torch that is passed from generation to generation, a flame that burns brighter with each new discovery. The everyday science that surrounds us is an open invitation to keep exploring, keep questioning, and keep discovering. For the universe is a treasure trove of mysteries waiting to be unraveled, and the journey of science is a lifelong adventure that beckons us to continue the exploration, embrace the wonder, and revel in the unending quest for knowledge.

The End.